To Ariel:
  My international partner
in climate change research and
good friend.

The Impact of Climate Change on Regional Systems

NEW HORIZONS IN ENVIRONMENTAL ECONOMICS

Series Editors: Wallace E. Oates, *Professor of Economics, University of Maryland, USA* and Henk Folmer, *Professor of General Economics, Wageningen University and Professor of Environmental Economics, Tilburg University, The Netherlands*

This important series is designed to make a significant contribution to the development of the principles and practices of environmental economics. It includes both theoretical and empirical work. International in scope, it addresses issues of current and future concern in both East and West and in developed and developing countries.

The main purpose of the series is to create a forum for the publication of high quality work and to show how economic analysis can make a contribution to understanding and resolving the environmental problems confronting the world in the twenty-first century.

Recent titles in the series include:

The Greening of Markets
Product Competition, Pollution and Policy Making in a Duopoly
Michael Kuhn

Managing Wetlands for Private and Social Good
Theory, Policy and Cases from Australia
Stuart M. Whitten and Jeff Bennett

Amenities and Rural Development
Theory, Methods and Public Policy
Edited by Gary Paul Green, Steven C. Deller and David W. Marcouiller

The Evolution of Markets for Water
Theory and Practice in Australia
Edited by Jeff Bennett

Integrated Assessment and Management of Public Resources
Edited by Joseph C. Cooper, Federico Perali and Marcella Veronesi

Climate Change and the Economics of the World's Fisheries
Examples of Small Pelagic Stocks
Edited by Rögnvaldur Hannesson, Manuel Barange and Samuel F. Herrick Jr

The Theory and Practice of Environmental and Resource Economics
Essays in Honour of Karl-Gustaf Löfgren
Edited by Thomas Aronsson, Roger Axelsson and Runar Brännlund

The International Yearbook of Environmental and Resource Economics
2006/2007
A Survey of Current Issues
Edited by Tom Tietenberg and Henk Folmer

Choice Modelling and the Transfer of Environmental Values
Edited by John Rolfe and Jeff Bennett

The Impact of Climate Change on Regional Systems
A Comprehensive Analysis of California
Edited by Joel B. Smith and Robert Mendelsohn

The Impact of Climate Change on Regional Systems

A Comprehensive Analysis of California

Edited by

Joel B. Smith

Stratus Consulting Inc., USA

Robert Mendelsohn

School of Forestry and Environmental Studies, Yale University, USA

NEW HORIZONS IN ENVIRONMENTAL ECONOMICS
In Association with the Electric Power Research Institute (EPRI)

Edward Elgar
Cheltenham, UK • Northampton, MA, USA

Published by
Edward Elgar Publishing Limited
Glensanda House
Montpellier Parade
Cheltenham
Glos GL50 1UA
UK

Edward Elgar Publishing, Inc.
William Pratt House
9 Dewey Court
Northampton
Massachusetts 01060
USA

A catalogue record for this book
is available from the British Library

Library of Congress Cataloguing in Publication Data

The impact of climate change on regional systems: a comprehensive analysis
    of California/edited by Joel B. Smith, Robert Mendelsohn.
        p. cm. — (New horizons in environmental economics)
    Includes bibliographical references and index.
        1. Climate changes—Environmental aspects—California. 2. Climatic
changes—Risk assessment—California. 3. Environmental management—
California. I. Smith, Joel B. II. Mendelsohn, Robert O., 1952–   . III. Series.
QC981.8.C5I4485   2006
551.69794—dc22

                                                                    2005037392

ISBN-13: 978 1 84542 747 4
ISBN-10: 1 84542 747 5

Printed and bound in Great Britain by MPG Books Ltd, Bodmin, Cornwall

Contents

v

Plates

Figures

Tables

Contributors

Richard M. Adams, Agricultural and Resource Economics, Oregon State University, Corvallis, OR.

Kathy E. Bashford, California Water Resources Research and Applications Center, Lawrence Berkeley National Laboratory, University of California, Berkeley, CA.

Raymond Drapek, Pacific Northwest Research Station, US Forest Service, Corvallis, OR.

Hector Galbraith, Galbraith Environmental Sciences, LLC, Newfane, VT.

Chuck Hakkarinen, Consultant, Belmont, CA.

Laurie L. Houston, Agricultural and Resource Economics, Oregon State University, Corvallis, OR.

Richard Howitt, Department of Agricultural and Resource Economics, University of California, Davis, CA.

Daniel Hudgens, Industrial Economics, Inc., Cambridge, MA.

Marion W. Jenkins, Department of Civil and Environmental Engineering, University of California, Davis, CA.

Russell Jones, Stratus Consulting Inc., Boulder, CO.

John D. Landis, Department of City and Regional Planning, University of California, Berkeley, CA.

James M. Lenihan, Pacific Northwest Research Station, US Forest Service, Corvallis, OR.

Jay R. Lund, Department of Civil and Environmental Engineering, University of California, Davis, CA.

Robert Mendelsohn, School of Forestry and Environmental Studies, Yale University, New Haven, CT.

Norman L. Miller, California Water Resources Research and Applications Center, Lawrence Berkeley National Laboratory, University of California, Berkeley, CA.

Ronald Neilson, Pacific Northwest Research Station, US Forest Service, Corvallis, OR.

James Neumann, Industrial Economics, Inc., Cambridge, MA.

Elizabeth Pienaar, Department of Agricultural and Resource Economics, University of California, Davis, CA.

Michael Reilly, Department of City and Regional Planning, University of California, Berkeley, CA.

Joel B. Smith, Stratus Consulting Inc., Boulder, CO.

Eric Strem, California Nevada River Forecast Center, National Weather Service, National Oceanic and Atmospheric Administration, Sacramento, CA.

Stacy K. Tanaka, Department of Civil and Environmental Engineering, University of California, Davis, CA.

JunJie Wu, Agricultural and Resource Economics, Oregon State University, Corvallis, OR.

Tingju Zhu, Department of Civil and Environmental Engineering, University of California, Davis, CA.

Preface

There is ever-increasing evidence that the burning of fossil fuels is causing global warming. As society faces difficult choices concerning what to do on mitigation and adaptation, it is ever more important to understand the likely future impacts of climate change. Impact studies need to be made at every scale both to understand the scope of the problem and also the appropriate response. Although there have been a number of national empirical studies of climate change impacts, there have been relatively few comprehensive and integrated regional impact studies. This book develops a thorough methodology for measuring regional climate impacts and applies that methodology to study California.

The study makes several important innovations over earlier regional as well as national climate impact studies. First the study develops detailed projections about population growth, land use, economic growth, and technological improvements over the twenty-first century. These projections are integrated into the assessments of future climate change impacts. Second, the study develops new methods that integrate the results of natural science models into economic modeling of agriculture, timber, and water. Third, the study includes a spatially detailed analysis of how climate might affect valuable habitats. Fourth, the study considers the potential for adaptation to respond to climate change impacts. While some of these have been done before, this book is the first study to integrate all these factors at the regional scale.

The study applies these new quantitative methods to estimate climate change impacts on California. The study was designed to develop consistent measures of each sector across the entire state. The analysis uses a wide range of climate change scenarios, ranging from warmer and much wetter to warmer and much drier to capture some of the key uncertainties about climate change at the regional scale. Because irrigation is an important factor in Californian agriculture, the farming model is integrated with the water model. This is an important advance in modeling irrigated agriculture. The study also makes an important advance in conservation planning. The study carefully overlays urban development with changes in ecosystems from different climate scenarios to determine safe strategies for protecting endangered species, in this case, coastal sage brush. The study examines the potential for adaptation to mitigate adverse impacts of

xv

climate change. While it is uncertain how we will adapt to climate change, there is no doubt society will attempt to adapt and no study of climate change impacts is complete without considering the potential effectiveness of adaptation. Finally, the study is careful to deal with analytical issues peculiar to regional studies such as the fact that regions are often small compared to markets.

Acknowledgments

The research was initiated and primarily funded through the California Energy Commission's (Commission) Public Interest Energy Research (PIER) program. The Electric Power Research Institute (EPRI) provided significant cofunding and assisted in research management. The views expressed in this book, however, are the author's and cannot be ascribed to these institutions. We personally wish to thank the following people who provided us invaluable help in pulling this book together. Christina Thomas edited the volume. Shiela DeMars helped organize the project. Diane Callow and Erin Miles formatted all the material and prepared a final manuscript.

Abbreviations

BLUE	best linear unbiased estimator
CALFED	California and Federal Program to manage California's water and aquatic ecosystems
CALVIN	California Value Integrated Network
CENTURY	vegetation simulation model
CFMMP	California Farmland Mapping and Modeling Program
CSS	coastal sage scrub
CVI	coastal vulnerability index
CERES	crop simulation model
CO_2	carbon dioxide
DP	dynamic programming
EIA	US Department of Energy, Energy Information Administration
DOF	Department of Finance
ENSO	El Niño/Southern Oscillation
EPIC	crop simulation model
EPRI	Electric Power Research Institute
ET	evapotranspiration
FEMA	Federal Emergency Management Agency
GAP	California Gap Analysis Project
GCM	general circulation model
GIS	geographic information system
GDP	gross domestic product
Hadley	Hadley general circulation model run HadCM2
IPCC	Intergovernmental Panel on Climate Change
LA	Los Angeles
MAP	mean area precipitation
MAPSS	vegetation simulation model
MAT	mean area temperature
MBTU	Million British Thermal Units
MC1	a dynamic vegetation model that estimates vegetation, carbon, nutrients, and water
MSA	metropolitan statistical area
NASS	National Agricultural Statistical Service
NCAR	National Center for Atmospheric Research

NOAA	National Oceanic Atmospheric Administration
NPP	net primary productivity
NWI	National Wetland Inventory
NWS	National Weather Service
OLS	ordinary least squares
PCM	Parallel Climate General Circulation Model
PDO	Pacific Decadal Oscillation
PIER	Public Interest Energy Research Program
RCM	regional climate model
S&LI	state and locally important
SAC-SMA	Sacramento Soil Moisture Accounting
SWAP	Statewide Water and Agricultural Production model
SWE	snow water equivalent
TAF	thousand acre-feet
UIUC	University of Illinois at Urbana-Champaign General Circulation Model
USDA	United States Department of Agriculture
USFS	United States Forest Service
USFWS	United States Fish and Wildlife Service
USGS	United States Geological Service
WRCC	Western Regional Climate Center

We wish to dedicate this book to our wives, Sarah Larson and Susan Mendelsohn. They put up with so much with our travel and our long hours working on projects such as this. We couldn't have done this book without their love and support. For that, we will always be grateful.

1. Introduction

Joel B. Smith and Robert Mendelsohn

Greenhouse gases are accumulating in the atmosphere as a result of burning fossil fuels and other human activities such as deforestation (Houghton et al., 2001). Without abatement, future emissions are expected to increase concentrations in the atmosphere substantially. These greenhouse gases essentially trap heat in the atmosphere. Climate scientists have concluded that recent increases in global temperatures are largely the result of increased greenhouse gas concentrations and that the continued increase in these concentrations will cause future climate warming (Houghton et al., 2001). Even if aggressive steps are taken to control the emissions of greenhouse gases, there is every reason to believe that the climate will continue to change, although the precise increases in temperature and changes in precipitation are uncertain. These changes could have substantial impacts on the Earth's coastal resources, water supplies, agricultural output, biodiversity, and other sectors. Every region around the world consequently needs to plan how they will adapt to climate change. This includes examining how society and nature may be affected by changes in climate and whether it is prudent to take actions now and in the future to avoid harmful impacts and take advantage of beneficial impacts.

This book is a comprehensive and integrated study of a single region, California. The state is important not just because it is home to 35 million people and has a large and vibrant economy, but because its development was the result of major engineering of its natural resources. In particular, the state has one of the most sophisticated water collection and distribution systems in the world. Table 1.1 gives a brief overview of California.

The project was designed to help California natural resource managers and other policy makers better understand the potential effects of climate change on the state. Indeed, the ultimate goal of the book is to quantify the impacts of climate change for California and make available information that natural resource policy makers can use to develop adaptive policies.

The book focuses on two key questions: To what extent would a change in climate affect California? And, what is the potential for the state to adapt to climate change?

1

Table 1.1   California statistics

Topography	423 970 km^2 (163 969 square miles). Third largest state in US. Elevations: from 86 m (282 ft) below sea level (Death Valley) to 4414 m (14 494 ft) atop Mt. Whitney. Greatest complexity of topography among US states. Mountain ranges: Klamath and Cascade in the north; Sierra Nevada along most of the eastern border; Coastal range close to the coast. Central Valley is between Sierra and Coastal Range. Major rivers: Sacramento, San Joaquin.
Climate	Varies widely. Precipitation: mainly in winter, heavy in northwestern portion; very dry inland and south of Los Angeles. Temperature: mild in maritime region and in winter, except in mountainous areas; very hot inland.
Ecosystems	Great diversity due to variability in elevation, topography, soils, climate: alpine tundra, temperate rainforest, intertidal wetlands, grasslands, arid deserts. More than 5000 native plants, 1000 vertebrates.[a] 48% of species endemic.[b] Forests cover 40% of land area.
Population	35.5 million (2003). Largest cities: Los Angeles, San Diego, San Jose, San Francisco, Long Beach, Fresno, Sacramento, Oakland. One-half of state's population in southern California.[c]
Economy	Largest in US; GSP of $1.4 trillion (2003).[d] Main economy sectors: agriculture, mining, entertainment, aerospace, government, education, tourism, high tech.
Agriculture	Largest value producer in US. 87 500 farms, average size 151 ha (374 acres). 4% of total farms in US, 13% of national farm income. 200 commercial crops, annual export value in late 1990s $6.6 billion (2% of total GSP).
Forestry	Timber harvest 3.8 million m^3 (1.6 billion board feet) in early 2000s, value of $570 million. Production of mainly softwoods, in coastal ranges and Sierra Nevada.

Notes:
a. Wilkinson (2002).
b. Schoenherr (1992).
c. US Census Bureau (2005).
d. US Bureau of Economic Analysis (2005).

This research builds on a number of previous studies of climate change impacts on California, including Smith and Tirpak (1989),[1] Knox and Scheuring (1991), Field et al. (1999), Wilkinson (2002), Barnett et al. (2004), and Hayoe et al. (2004). These previous studies addressed impacts of climate change on such resources as water supplies, agriculture, coastal resources, air quality, and natural resources.

This study makes several important improvements over earlier regional and national assessments of climate change impacts. First, it develops two relatively detailed projections about population growth, land use, economic growth, and technological improvements in California over the twenty-first century. The scenarios are discussed in more detail below.

Second, the study integrates the results of natural science and economics. It does not estimate just physical changes such as change in runoff or snowpack (for example Hayhoe et al., 2004) but also the consequences of those changes and the potential ability of societal systems to adapt to them. For example, the study develops a complex, integrated, water–agriculture economic model to estimate changes in water supply, demand, and distribution. This model predicts outcomes in both the water sector and the agricultural sector. It is by far the most detailed and integrated analysis of water resources and irrigated agriculture yet conducted.

Third, the study includes a detailed analysis of how California's terrestrial ecosystems and biodiversity may be affected by climate change. It includes a spatially detailed analysis of how climate change might change valuable habitats. It then combines that analysis with a highly detailed set of projections on future land development to examine the combined impacts of climate change and development on the state's biodiversity.

The structure of the study is displayed in Figure 1.1. The first step was creation of baseline scenarios of socio-economic conditions (population, income, and land use); these baseline socio-economic scenarios are discussed in Chapter 2. The baselines are dynamic projections of growth in the region. Although it is likely that California will continue to grow, it is not clear how quickly it will do so. To capture some of the uncertainty about future population and economic growth, the study examines two dynamic pathways, a low-growth and a high-growth scenario. In the low-growth scenario, 67 million people are living in California by 2100 and in the high-growth scenario there are 92 million people by 2100. Both socio-economic scenarios assume that per capita income will grow, although at different rates.

Chapter 3 examines how land use in the state could change over time in response to the population growth. The study addresses the extent to which infill development can provide sufficient housing and the extent to which new development may be needed. Further, the land use modeling predicts what counties are likely to see the highest growth in population. These

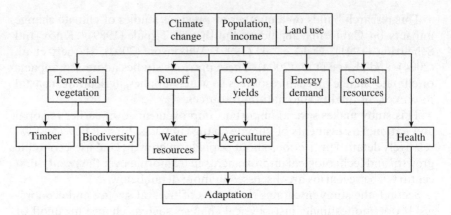

Figure 1.1   Study design

assumptions turn out to be quite important for understanding what could happen to natural open space (Chapter 6) and what could happen to agricultural land (Chapter 11) under the combined effects of population growth and climate change.

The second step was to create a set of climate change scenarios that reflect the broad potential array of future changes in climate (Chapter 4). The Intergovernmental Panel on Climate Change (IPCC) has determined that the range of possible temperature changes for the world by 2100 lies between 1.4°C and 5.8°C (Houghton et al., 2001). While there is little doubt that temperatures in California will increase over the century, there is uncertainty about whether annual precipitation will increase or decrease. To be useful for policy makers, it is important that climate change scenarios reflect this broad range of potential changes in climate for California (or any region being studied). Some studies have used only climate change scenarios with substantial increases in precipitation (for example Wilkinson, 2002), while others have relied mainly on scenarios that estimate substantial decreases in precipitation (for example Hayhoe et al., 2004). Relying on limited scenarios can be misleading and give the wrong impression that a single scenario is the only possible outcome for California over the coming century. We selected several temperature and precipitation scenarios to capture a broad range of possible outcomes. All the scenarios used in this study include higher temperatures, but some include increased precipitation, others include little change in precipitation, and others include decreased precipitation. By including both warmer and wetter, and warmer and drier scenarios, the study reflects the uncertainty about potential changes in California's climate and presents a broader range of potential impacts than would come from a more limited set of climate change scenarios.

This book provides an easily accessible overview of results. More detailed results are available on the Commission's website: http://www. energy.ca.gov/pier/final_project_reports/500–03–058cf.html.

The scenarios were then used in individual studies displayed in the boxes in the second and third rows of Figure 1.1 to assess potential impacts of climate change on various sectors of the economy and the natural environment. The results of some of these studies were used as inputs into other related studies.

Chapter 5 estimates how the distribution of major terrestrial ecosystems could be affected by climate change. It also estimates how productivity of the ecosystems could change and how fire frequency could be affected by climate change. The results of Chapter 5 on terrestrial vegetation changes were used in both Chapter 6 on biodiversity and Chapter 7 on forestry. Chapter 6 combines the projections of land use change with the estimates of changes in terrestrial ecosystem location. The model provides one of the first estimates of dynamic ecosystem adjustments in the literature. This analysis is used to examine how biodiversity may change and how one critical habitat, coastal sage scrub, could be affected by development and climate change. Chapter 7 examines how changes in the productivity and location of species could affect the productivity of the forestry sector in California. The chapter combines estimation of how climate change might affect softwood growth in the state with projections of global forestry production changes that might affect timber prices.

Chapters 8 through 11 present the most detailed and integrated assessment conducted of how climate change could affect California's water supplies and agricultural economy. Chapter 8 estimates changes in runoff in six representative and important river basins in California. Chapter 9 estimates how crop yields could be affected by continued but slow improvements in technology as well as changes in climate. The latter considers the direct effects of changes in temperature and precipitation and the carbon dioxide fertilization effect. The study estimates changes in crop yields and in demand for irrigation.

The results of Chapter 8 on runoff and Chapter 9 on crop water were used in both Chapter 10 on water resources and Chapter 11 on agriculture. Chapters 10 and 11 use general equilibrium modeling to examine climate impacts on the water and agricultural systems, respectively. These two chapters are tightly linked because the vast majority of the state's water supplies are used by irrigated agriculture. The water resources chapter uses a complex model where changes in snowpack, surface water runoff, and groundwater supplies are inputs. The water model estimates changes in water demand from population growth and warming. The model then calculates where water supplies would be most efficiently allocated. The agriculture model starts with changes in land use, water supplies, crop yields, and irrigation demand from climate change. The model then calculates how many acres farmers will plant and irrigate, where they would be planted (and not planted), and which crops would be chosen. The result of all these adaptations leads to a final welfare estimate of agricultural impacts.

The remaining chapters address two other key sectors, energy and coastal resources. Chapter 12 examines the potential effect of climate change on energy demand for space cooling and heating. Chapter 13 examines what might happen along California's developed coast if the sea level rises. It estimates whether developed coastlines would be protected from sea level rise or be inundated.

The book concludes with Chapter 14 which summarizes the overall results. Market impacts are summed and compared. Non-market impacts are discussed as quantitatively as possible. The strengths and weaknesses of the overall methods are reviewed and future research initiatives are proposed. The conclusion also discusses the many policy implications of the results. How important are climate change impacts to California? What policy changes are needed to help the state adapt to future scenarios? How soon should planners prepare for climate change?

This study can be a model for how other regional assessments can be conducted. In particular, the use of multiple scenarios of socio-economic development, a wide range of climate change scenarios, and detailed and integrated analysis of impacts and adaptation potential are important for better understanding the potential consequences of climate change.

NOTE

1.   Smith and Tirpak (1989) had a chapter on California, whereas the other studies focused exclusively on California.

REFERENCES

Barnett, T., R. Malone, W. Pennell, D. Stammer, A. Semtner, and W. Washington. 2004. The effects of climate change on water resources in the West. *Climatic Change* **62**: 1–11.

Field, C.B., G.C. Daily, F.W. Davis, S. Gaines, P.A. Matson, J. Melack, and N.L. Miller. 1999. *Confronting Climate Change in California: Ecological Impacts on the Golden State*. Union of Concerned Scientists, Cambridge, MA.

Hayhoe, K. and 18 co-authors. 2004. Emissions pathways, climate change, and impacts on California. *Proceedings of the National Academy of Sciences* **101**: 12422–7.

Houghton, J., Y. Ding, D. Griggs, M. Noguer, P. van der Linden, X. Dai, K. Maskell, and C. Johnson (eds). 2001. *Climate Change 2001: The Scientific Basis.* Third Assessment Report of the Intergovernmental Panel on Climate Change, Cambridge University Press, Cambridge, UK.

Knox, J.B. and A.F. Scheuring (eds). 1991. *Global Climate Change and California: Potential Impacts and Responses.* University of California Press, Berkeley, CA.

Schoenherr, A.A. 1992. *A Natural History of California.* University of California Press, Berkeley, CA.

Smith, J.B. and D. Tirpak (eds). 1989. *The Potential Effects of Global Climate Change on the United States.* EPA-230-05-89-050. US. Environmental Protection Agency, Washington, DC.

US Bureau of Economic Analysis. 2005. Regional Economic Accounts. http://www. bea.doc.gov/bea/regional/gsp/action.cfm. Accessed 27 January 2005.

US Census Bureau. 2005. *Statistical Abstract of the United States.* http://www. census.gov/prod/www/statistical-abstract-04.html. Accessed 19 January 2005.

Wilkinson, R. 2002. *Preparing for a Climate Change: The Potential Consequences of Climate Variability and Change – California.* A Report of the California Regional Assessment Group. National Center for Geographical Information Analysis, University of California, Santa Barbara, CA.

2. Socio-economic changes

Joel B. Smith

2.1 INTRODUCTION

This study estimates the impacts of climate change in 2020, 2060, and 2100 in California. Because these climate impacts would occur well into the future, it is important to consider what the state might look like in socio-economic terms by these dates.[1] California experienced tremendous socio-economic change in the twentieth century. Its population grew from 1.5 million in 1900 to 35.5 million in 2004 (US Census Bureau, 2005). While the rate of growth in the twenty-first century might not match that of the twentieth century, the absolute increase in population may be greater. Further, California is likely to become richer and have vastly different technologies. Consequently, California in 2100 may be quite different from California today.

These baseline changes in California's population, level of development, economy, and environment could dramatically change the state's sensitivity to climate change. More people will be exposed to the impacts of climate change, but they are also likely to have more financial resources. More income is predicted to increase energy demand, which would make that sector more sensitive to higher temperatures. On the other hand, people will have more technical resources, and the health sector could be less affected. For example, there may be more heat waves, but the use of air conditioning is projected to increase, reducing the time people spend in hotter climate conditions. Continued improvements in agricultural yields from new technologies are projected as well. By increasing total production, the agriculture sector becomes larger and therefore more vulnerable to climate change, but new crop varieties developed for heat may be less sensitive to warming than current varieties.

Projecting future immigration, demography, economic production (for the state, for the nation, or for the world), technology, urbanization, transportation, pollution, and demand for environmental services is not easy. Changes in sectors that will be sensitive to climate change, such as water resources, agriculture, timber, recreation, and coastal resources are difficult to foresee. It is hard to predict how much water demand, agricultural

Table 2.1   Growth scenarios for California

Variable	Year			
	2000	2020	2060	2100
Low population growth scenario				
Population (million)	35	45	67	67
Income per capita (000, $)	40	54	98	177
High population growth scenario				
Population (million)	35	45	70	92
Income per capita (000, $)	40	44	54	66

productivity, timber demand, recreational use, and coastal development there will be over 100 years. Any prediction of future socio-economic changes will be uncertain.

In this analysis, we try to test how important the baseline might be to climate impacts by developing two scenarios of changes in socio-economic baseline conditions. With each baseline scenario, we compare climate change scenarios against current climate conditions. By comparing the outcomes between the different baseline scenarios, one can begin to see what role the baselines themselves have in affecting the vulnerability of the state to climate change. Of course, it is not possible to capture all the uncertainties in only two scenarios. The scenarios do not reflect all the possible future conditions in California. Their main function is to help us understand how the state's sensitivity to climate change might change over time and how change in some key socio-economic factors may affect that sensitivity.

There are tremendous uncertainties about the future development of California. The two scenarios of baseline changes were developed to capture a range of outcomes that might be important to climate change impacts (Table 2.1). The scenarios do not dictate specific sectoral changes; instead, sector-specific assumptions and sensitivity studies were left to the authors of the sector studies to decide. The analysts for each sector were asked to make sectoral assumptions that would be consistent with the statewide socio-economic projections in this chapter.

These scenarios focus on the most important future parameters, population growth rates and the growth in per capita income in California.

2.2   POPULATION

Population forecasts for California are particularly difficult to make because they combine demographic changes in the existing population with net migration. It is relatively easy to project how births and deaths may change in the near future because they are largely determined by the existing population's age distribution. However, the further into the future the projection, the more important the future birth and death rates become. If these rates change dramatically from current experience, the future population could look quite different by 2060, and vastly different by 2100. Projecting changes in net migration over time is also complicated. If future US income continues to outstrip by far the per capita income of Latin America or other developing parts of the world, there may be tremendous pressure for net migration into California. Under this scenario, the state would certainly continue to grow. In contrast, if Latin America or the other regions develop quickly, this pressure may evaporate and the state's population could stabilize.

To capture these two dramatic population situations, high and low population growth scenarios were developed. The high population growth scenario assumes that California's population, which has grown by 500 000 people per year (5 million per decade) since World War II, would continue to grow at a rate slightly higher than that through the next century. In this scenario, the state population would expand to 92 million by 2100. The low population growth scenario projects that this growth path would slow over time and eventually stabilize at 67 million. In the low scenario, the population would reach 45 million by 2020, and then gradually approach 67 million by 2060, when it would stabilize. These two scenarios were designed to test whether the size of the state's future population is important in determining California's climate sensitivity. Readers should not overinterpret the assumptions and assume that these are the only two possible futures for California or that these changes capture the full possible range of future population growth. However, these population projections are consistent with estimates made by others in the state (for example Johnson, 1999).[2]

2.3   ECONOMIC GROWTH

We assumed that lower population growth would result in higher growth of per capita income and higher population growth would result in lower growth of per capita income. This is because we assumed higher population growth would result in relatively large immigration of unskilled labor. Historical rates of growth during the last century averaged 2 percent in the United States. However, it is not clear if this exponential rate of growth can be maintained indefinitely. We have consequently assumed somewhat lower growth rates in the twenty-first century. We assumed a 1.5 percent growth rate for California per capita income in the low population growth scenario. However, if there is a large influx of unskilled immigrants, per capita income growth is likely to lag behind this growth rate. Under the high population growth scenario, we assume that income would grow at only 0.5 percent per year. As can be seen in Table 2.1, these scenarios capture a wide range of potential future socio-economic conditions.

2.4 CONCLUSIONS

The socio-economic scenarios were created not as a prediction of California's future population growth and development but to consider the impacts of climate change against a background of a wide range of plausible socio-economic futures. The purpose of the scenarios is to help understand how different development paths would affect the impacts of climate change. While the socio-economic scenarios reflect a wide range of future conditions for California, other changes that could result in different sensitivities to climate change are also possible. For example, new technologies may come along that either are very climate sensitive, or reduce the sensitivity of existing sectors to future climates.

The socio-economic scenarios were used to derive changes in urbanization (Chapter 3). The population scenarios were used in all of the impact studies examining climate change, including timber (Chapter 7), water resources (Chapter 10), agriculture (Chapter 11), energy (Chapter 12), and coastal resources (Chapter 13). The sectoral studies made varying assumptions about income per capita changes.

NOTES

1. This chapter is based on results of a workshop with experts from the California impact team, agency representatives from the State of California, and researchers from the US Department of Energy (DOE) laboratories, universities, and research institutions in California in Sacramento on 12 June 2000. The workshop addressed the baseline scenarios that would be used in the global warming study of California. The report from the workshop is in Appendix II of the report to the California Energy Commission (available at http://www.energy.ca.gov/reports/2003-10-31_500-03-058CF_A02.PDF).
2. Johnson (1999) cites published projections of California's population by 2050. Those projections range from 50 million to 70 million. We assumed that the high projection would continue at the same rate and that the low projection would level off. While the population scenarios used in this study are a broad range, the full potential range of California's population in 2100 is even broader.

REFERENCES

Johnson, H. 1999. How Many Californians? *California Counts* **1**. Public Policy
    Institute of California, San Francisco, CA. Available at http://www.ppic.org/
    content/pubs/cacounts/CC_1099HJCC.pdf.
US Census Bureau. 2005. *Statistical Abstract of the United States.* http://www.
    census.gov/prod/www/statistical-abstract-04.html. Accessed 19 January 2005.

3. Urbanization scenarios

John D. Landis and Michael Reilly

3.1 INTRODUCTION

By 2020, most forecasters agree, California will be home to between 43 and 46 million residents – up from 35 million in 2003. Beyond 2020 the size of California's population is less certain. Depending on the composition of the population, and future fertility and migration rates, California's 2050 population could be as little as 50 million or as much as 70 million. One hundred years from now, if present trends continue, California could conceivably have as many as 90 million residents (see Chapter 2 for a discussion of the population scenarios for California used in this study).

Where these future residents will live and work is unclear. For most of the twentieth century, two-thirds of Californians have lived south of the Tehachapi Mountains and west of the San Jacinto Mountains – in that part of the state commonly referred to as southern California. Most of coastal southern California is already highly urbanized, and relatively little vacant land is available for new development. More recently, slow-growth policies in northern California and declining developable land supplies in southern California are squeezing ever more of the state's population growth into the San Joaquin Valley.

How future Californians will occupy the landscape is also unclear. During the last 50 years, the state's population has grown increasingly urban. Today, nearly 95 percent of Californians live in metropolitan areas, mostly at densities of fewer than ten individuals per acre. Recent growth patterns have strongly favored locations near freeways, most of which were built in the 1950s and 1960s. With few new freeways on the planning horizon, how will California's future growth organize itself in space? By national standards, California's large urban areas are already reasonably dense, and economic theory suggests that densities should increase further as California's urban regions continue to grow. In practice, densities have been rising in some urban counties, but falling in others.

This chapter presents the results of a series of baseline population and urban growth projections for California's 38 urban counties through 2100. The methods section (section 3.2) outlines the approach and data used to develop the various projections. The third section (Section 3.3) presents the baseline scenario, which is the projection of growth into the future. A final section (section 3.4), on baseline impacts, quantitatively assesses the consequences of the baseline projections on wetland, farmland, and habitat loss.

3.2   METHOD

Developing short-term forecasts in a state as diverse and fluid as California is a difficult proposition. Developing long-term forecasts, whether for 20, 50, or 100 years, is harder still. Developing long-term forecasts that are spatially explicit – that project which lands are likely to be developed and which are not – is closer to art than science.

At a conceptual level, our forecasting method is actually quite simple (Figure 3.1; Landis et al., 1998).[1] We begin by calibrating a spatial-statistical model of historical development patterns spanning 1988 to 1998 (Figure 3.1, Step A). The calibrated model parameters are then used with contemporary spatial data to generate a development probability surface describing the likelihood that particular undeveloped sites will subsequently be developed (Figure 3.1, Step B). This is the 'where' part of the equation. *When* development happens is a function of state and county population growth pressures (Figure 3.1, Step 1), the share of population accommodated through infill development (Figure 3.1, Step 2), and the density at which development occurs (Figure 3.1, Step 3). Projected population growth, net of infill, is then allocated to allowable development sites in order of their projected development probability (from Step B) at a designated development density (Figure 3.1, Step 4). Once a future allocation has been completed (for example for the 2000–2020 period), infill rates, densities, and development probabilities are updated to reflect any intervening changes. The model is then run again (Steps 1 through 5) for subsequent periods.

3.2.1   Growth Model Calibration

Before a statistical model can be used to generate future projections, it must be calibrated. With non-spatial models, this usually involves fitting a line or curve to historical data. With spatial data, equations and parameters that are sensitive to locational and non-locational influences are developed and estimated. In this case, the model being calibrated relates changes in the development status of particular sites between 1988 and 1998 – measured as a matrix of 1 ha grid cells – and their various physical, locational, and

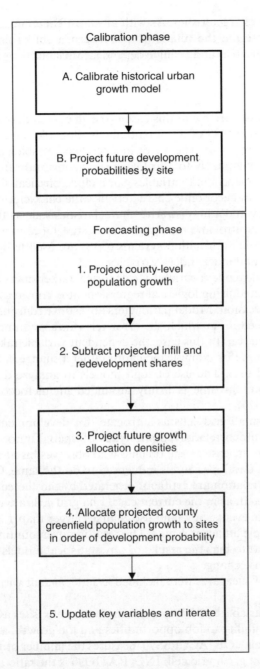

Figure 3.1   Urban growth forecasting process

administrative characteristics. As with all statistical models, the estimated parameters describe the relationship between a set of independent or explanatory variables and a single dependent variable:

$$\text{Prob [undeveloped site } 0 > \text{developed]} = f(X1\ 0,\ X2\ 0,\ \dots\ Xn\ 0)\quad(3.1)$$

The dependent variable in this case is the change in development status between 1988 and 1998 of all potentially developable sites, measured as a matrix of 1 ha grid cells. The Xs, or independent variables, are those attributes thought most likely to affect each site's conversion from non-urban to urban use. Independent variables can include physical site characteristics, locational and economic characteristics, the characteristics of nearby sites, and policy and administrative characteristics such as the presence of a local growth control and management measure. Once variables are measured, geographic information systems (GIS) are used to spatially match the dependent and independent variables.

Because the dependent variable is categorical rather than continuous, the model is estimated using logistical regression, also known as 'logit', rather than linear regression. Model parameters are estimated using a maximum-likelihood procedure in which the error terms are presumed to follow a Weibull distribution. In this case, the dependent variable takes on just two categorical values (for example indicating either a change in land use or no change in land use). The use of logit models to analyze discrete choices at a single point in time is firmly grounded in microeconomic theory (McFadden, 1974).

The use of small grid cells as surrogates for development sites exacerbates spatial autocorrelation. Some types of spatial autocorrelation are legitimate, as in the case of the rancher who observes his next-door neighbor selling to a developer and is influenced to do the same. Other types of spatial autocorrelation are artifacts, generated by the choice of the spatial unit of analysis. If, as in the current case, 1 ha grid cells are used to record land use change events, any land use changes larger than 1 ha will be recorded as multiple adjacent events. The resulting overcounting of land use change will tend to bias the results of any statistical models calibrated on the basis of those changes.

Four types of measures were included as independent variables:

1. Demand variables, which measure the demand for sites as a function of their accessibility to job opportunities and job growth, as well as local income levels. JOB_ACCESS90 measures the number of jobs within 90 minutes of a given grid cell. INC_RATIO90 is the ratio of community median household income to county median household income.

2. Own-site variables, which measure the physical and land use characteristics of each grid cell as determinants of its development potential. Four own-site variables are modeled: *FRWY_DISTSQ*, a measure of the squared distance from each site to the nearest freeway; *PRIME_FARM*, a dummy variable that indicates whether the California Farmland Mapping and Modeling Program (CFMMP) classifies the site as prime farmland; *SLOPE*, the average percentage slope of each site; and *FLOOD*, a dummy variable indicating whether the site falls within the 100-year flood zone designated by the Federal Emergency Management Agency (FEMA).[2]

3. Adjacency and neighborhood variables, which summarize the environmental and land use characteristics of adjacent and neighboring grid cells. Four neighborhood variables are modeled: *SLOPE_1KM*, the average slope of the cells within 1 km of each subject site; *SLOPE_2–3KM*, the average slope of sites within the 2 to 3 km ring around each subject site; *FLOOD_1KM*, the share of sites within 1 km of the subject site that are located in the FEMA 100-year flood zone; and *FLOOD_2–3KM*, the share of sites within the 2 to 3 km ring around each subject site.

4. Regulatory and administrative variables, which are intended to capture the development-encouraging or -constraining effects of different land use policies and regulations. *IN_CITY* denotes whether or not a site is located within an incorporated city. We expect sites located within incorporated cities to be more likely to be developed than unincorporated county lands. A second set of dummy variables, one for each county, is included to reflect intercounty differences in land use regulation.

The calibration sample consists of all 1 ha sites in a county that were undeveloped as of 1988, that were not publicly owned (and therefore could be developed), that had a slope of less than 15 percent, and that were within 15 km (9 miles) of a major highway or existing urban development.

Among southern California counties, the factors that most increased the likelihood of site development during the 1990s were freeway proximity (*FRWY_DISTSQ*), job accessibility (*JOB_ACCESS*), being located in a city (*IN_CITY*), and being located in Santa Barbara or San Diego counties. Steeply sloped sites were less likely to be developed than flatter sites, and prime farmlands were somewhat less likely to be developed. Reflecting Nimby ('not in my backyard') pressures, sites in upper-income communities were significantly less likely to be developed than sites in middle- or lower-income communities.

Among northern California counties, the factors that most increased the likelihood of site development during the 1990s were freeway proximity, being located in a city, and being located in Napa, Sonoma, Santa Cruz, Monterey, and Stanislaus counties. Compared to southern California, steeply sloped sites and prime farmlands in northern California were far less likely to be developed than flatter and less fertile locations. Accessibility to jobs, although a positive influence on development, was far less significant in northern California than southern California. Surprisingly, sites in wealthy communities in northern California were actually more likely to be developed than sites in poorer communities – the opposite situation of that in southern California.

Among Sacramento area counties, the factors that most affected the likelihood of site development during the 1990s were freeway proximity, whether the site was located in a flood zone (*FLOOD*), and whether the site was on prime farmland (*PRIME*). Sites near freeways were much more likely to have been developed; flood zone and prime farmland sites were much less likely to have been developed. Job accessibility was also an important influence. Sites located in incorporated cities were only marginally more likely to be developed than unincorporated sites – a finding in contrast to the southern and northern California regions, where development strongly favored incorporated sites. The effect of community income on development activity, although negative, was also slight.

Among counties in the southern San Joaquin Valley – including Fresno, Kern, Kings, Madera, Merced, and Tulare – the two factors that most affected the likelihood of site development during the 1990s were regional job accessibility and whether the site was located in an incorporated city. Sites with good accessibility to jobs were much more likely to have been developed, as were sites in incorporated cities. As in the Sacramento region, hillside sites were slightly more likely to have been developed than valley sites. Still, freeway accessibility had a much smaller effect on site developability in the southern San Joaquin Valley than elsewhere in the state. On the negative side, flood zone sites and sites located on prime farmland were less likely to have been developed than other, less environmentally sensitive sites, although the differences were not large.

Once estimated, the various model parameters can be used to generate development probability scores for all remaining undeveloped sites. Using these scores for forecasting assumes that the particular factors that influenced development in the recent past will continue to do so in the future, and in the same combination. To the extent that the future brings no large technological or land use policy changes, or significant shifts in household and business location preferences, the assumption that future land development trends will follow those of the past may be quite reasonable. On the other hand, to the extent that land use preferences, policies, and technologies all change, the usefulness of models calibrated using historical data is obviously reduced.

3.2.2   Patterns of Job Growth

Depending on the region, job accessibility is either the second, third, or fourth most important determinant of urban growth patterns in California. Most available employment projections are short term, and subject to constant revision. In terms of space, most job projections are undertaken at the metropolitan statistical area (MSA) or county level – for reasons of data availability as much as for modeling capability.

We start with the presumption that there is a more or less regular relationship between the size of a region's population and its employment base.[3] By accepting this assumption, we can use believable regional population projections as a starting point for developing serviceable regional employment projections.

For our purposes, the major challenge is not projecting the total number of new jobs. Instead, it is to figure out where in each region those new jobs are most likely to locate. Fortunately, the long-term spatial trend is quite clear. Broadly speaking, we expect jobs in California to continue their historical pattern of intrametropolitan decentralization. Before 1950, most basic jobs in the US economy were located in urban cores.[4] Since 1950, job growth has increasingly favored suburban communities over urban cores. Since 1980, almost all net basic job growth has occurred outside traditional central cities. First the Los Angeles region and more recently the San Francisco Bay Area have been national leaders in the trend toward increased job decentralization.

To project future job decentralization, we started by comparing the 1990 and 2000 spatial distribution of jobs in each California metropolitan region. Next, 10 km wide rings were generated outward from each regional center, and used to count the number of job centers and total number of jobs in each ring. Next, a spatial shift-share model was applied to decompose 1990–2000 city and CDP (census designated place) job changes into three components:

1. A regional growth component (RGC), calculated as the percent change in regional jobs between 1990 and 2000:

$$\text{(2000 regional jobs/1990 regional jobs)}. \tag{3.2}$$

2. A ring change component (RCC), calculated as the difference between the 1990–2000 percent change in jobs in each ring and RGC:

$$\text{(2000 ring jobs}_i\text{/1990 ring jobs}_i) - \text{(2000 regional jobs/1990 regional jobs)}. \tag{3.3}$$

Rings with RCC values greater than zero added jobs at a faster rate than the region as a whole. Rings with RCC values less than zero added jobs at a slower rate than the region as a whole.

3. A local change component (LCC), calculated as the difference between the 1990–2000 percent change in jobs in each city or CDP and the 1990–2000 percent change in jobs in its respective ring:

$$(2000 \text{ local jobs}_j/1990 \text{ local jobs}_j) - (2000 \text{ ring jobs}_i/1990 \text{ ring jobs}_i). \qquad (3.4)$$

Localities with LCC values greater than zero added jobs at a faster rate than their rings. Localities with LCC values less than zero added jobs at a slower rate than their rings.

Among the four regions we profiled, the rate of inner-ring job growth was greatest in northern California and the rate of outer-ring job growth was greatest in southern California.

3.2.3   Forecasting Procedures

As we noted previously, forecasting and scenario-building involves five distinct steps:

1. Project county-level population growth through 2100. County population projections for 2020 and 2040 were obtained from the California Department of Finance (DOF), Population Research Unit.[5] We used these projections to estimate annualized population growth rates (by county) spanning the periods 2000–2040 and 2020–40. Projected forward, these growth rates were used in turn to forecast county population totals for 2050 and 2100.

2. Subtract projected infill and redevelopment shares. A significant share of projected population growth will occur within the existing urban footprint in the form of infill or redevelopment. Infill shares tend to rise over time as remaining greenfield areas are used up and as developers reconsider infill lands that were previously passed over. A cross-sectional regression model was developed relating current county infill shares to remaining greenfield land supplies. We then used this model to project future infill and greenfield population shares for 2020, 2050, and 2100.

3. Project future allocation densities. The amount of greenfield land consumed by future population growth will depend both on the magnitude of growth and on its gross density. Marginal gross densities – the gross

densities of new development – were estimated for each county by dividing the change in the population between 1988 and 1998 by the change in urbanized land area for the same period. Theory suggests that densities should rise as available greenfield lands are used up, because developers seek to use remaining lands more intensely. We developed a cross-sectional regression model to relate 1988–98 marginal densities to remaining greenfield land supplies, then used this model to project future allocation densities by county for 2020, 2050, and 2100.

4. Allocate projected greenfield population growth to undeveloped sites in each region in order of development probability. Starting with the hectare-scale development probability scores derived above, a series of exclusion conditions is developed identifying sites that are to be precluded from development. Projected population growth (from Step 2) for 2000–2020 is then allocated to sites at projected densities (from Step 3) in order of development probability (from high to low) subject to any exclusion conditions.

5. Update key variables to reflect projected employment growth and allocated population growth. Steps 4 and 5 are iterated for 2020–50 and 2050–2100. Because of the analytical power of GIS, different forecasting steps can be undertaken at different spatial scales and then reconciled. Population growth, greenfield shares, and allocation densities, for example, are all identified and projected (Steps 1, 2, and 3) at the county level. Development probability scores, on the other hand, are estimated for individual 1 ha sites, accounting for differences among counties and regions. Employment projections, an input into the allocation procedure (Step 4), are developed for individual job centers. Distance to city boundaries, another input into the allocation procedure, is estimated and updated for incorporated cities.

The following sections explain and discuss each of these procedures in greater detail.

3.2.4 Population Projections: Huge Growth Ahead

The DOF's county-level population projections vary from a high of 3 percent per year for Imperial County to a low of –0.4 percent per year for San Francisco County.[6] Annualized 2020–40 growth rates are somewhat lower, and range from a high of 2.65 percent, also for Imperial County, to a low of –0.5 percent for San Francisco County.

High growth rates are rarely sustainable over the long term. Similarly, the growth rates of low-growth counties located in high-growth states tend to pick up over time. To better reflect this county–state convergence, we averaged each county's 2000–2040 and 2020–40 growth rates with those of the state as a whole.

Based first on the lower 2020–40 combined rate, and second on the higher 2000–2040 combined rate, we projected each county's population forward to 2050 and 2100. California's largest county, Los Angeles, is projected to grow from 10 million people in 2000 to 15.5 million by 2050. The populations of Riverside, San Bernardino, and San Diego counties are projected to exceed 5 million by 2050. The population of Orange County is estimated to grow from 2.8 million in 2000 to more than 4.5 million in 2050. Elsewhere, the 2050 population of the largest county in northern California, Santa Clara, is projected to be just under 3 million. With a 2050 population of 2.4 million, Sacramento County will be the most populous in the Central Valley. Added up, the total 2050 population of California's 58 counties would exceed 66 million.

Projecting further forward to 2100 presents additional challenges. Given the immense size of California's population, even the lower 2020–40 growth rates are likely to be unsustainable over time. To better reflect the natural tendency for growth rates to decline as the population increases, we reduced both the lower 2020–40 composite growth rate and the higher 2000–2040 composite growth rate by 50 percent before applying them to the 2050–2100 period.[7] Table 3.1 and Figure 3.2 present the final set of population projections for 2020, 2050, and 2100 organized by county.

Despite the imposed slowdown in growth rates, California's largest counties are projected to continue to grow. California's largest county, Los Angeles, is estimated to grow from 15.5 million people in 2050 to 20.4 million by 2100. The populations of Riverside, San Bernardino, and San Diego counties would each approach or exceed 5 million by 2050 and 7.5 million by 2100. The population of Orange County is projected to grow from 4.5 million in 2000 to nearly 5 million by 2050 and nearly 6 million by 2100. Elsewhere, the 2100 population of the largest county in northern California, Santa Clara, is estimated to be about 3.8 million. With a 2100 population of 3.3 million, Sacramento County would still be the most populous in the Central Valley. Added up, the total 2100 population of California's 58 counties could very well exceed 92 million.

The huge size of these projections – particularly among southern California counties – clearly indicates the dangers implicit in the long-term use of average annual growth rates. Even so, as large as these projections may seem, they are not unbelievable. California's population in 1900 was just over 1 million. One hundred years later, the state's population stood at nearly 35 million.

Table 3.1   Population projections for 2020, 2050, and 2100 by county

Major region	County	Sub-region	Population estimates and forecasts			
			2000 (source: DOF)	2020 (source: DOF)	2050[a]	2100[a]
Southern California	Los Angeles	Central	9 838 861	11 575 693	15 497 560	20 400 280
	Imperial	South	154 549	298 700	612 914	1 000 884
	Orange	South	2 833 190	3 431 869	4 535 936	5 932 517
	San Diego	South	2 943 001	3 917 001	5 831 574	8 097 302
	Subregional total		5 930 740	7 647 570	10 980 424	15 030 702
	Riverside	Inland Empire	1 570 885	2 773 431	5 335 081	8 431 480
	San Bernardino	Inland Empire	1 727 452	2 747 213	4 983 011	7 644 175
	Subregional total		3 298 337	5 520 644	10 318 093	16 075 656
	Santa Barbara	North	412 071	552 846	905 294	1 318 823
	Ventura	North	753 820	981 565	1 456 134	2 018 255
	Subregional total		1 165 891	1 534 411	2 361 429	3 337 078
	Regional total		20 233 829	26 278 318	39 157 506	54 843 715
Northern California	Alameda	Central	1 470 155	1 793 139	2 287 126	2 938 378
	Contra Costa	Central	931 946	1 104 725	1 394 436	1 782 151
	San Francisco	Central	792 049	750 904	710 034	785 565
	San Mateo	Central	747 061	855 506	1 044 065	1 312 014
	Santa Clara	Central	1 763 252	2 196 750	2 884 875	3 760 965
	Subregional total		5 704 463	6 701 024	8 320 538	10 579 072
	Marin	North	248 397	268 630	325 152	406 920
	Napa	North	127 084	157 878	214 934	285 317
	Solano	North	399 841	552 105	789 742	1 074 736
	Sonoma	North	459 258	614 173	845 837	1 129 343
	Subregional total		1 234 580	1 592 786	2 175 666	2 896 317
	Monterey	South	401 886	575 102	1 006 978	1 517 431
	San Benito	South	51 853	82 276	133 208	192 948
	San Luis Obispo	South	254 818	392 329	617 709	882 227
	Santa Cruz	South	260 248	367 196	572 017	812 597
	Sub-regional total		968 805	1 416 903	2 329 912	3 405 203

Table 3.1   (continued)

Major region	County	Sub-region	Population estimates and forecasts			
			2000 (source: DOF)	2020 (source: DOF)	2050[a]	2100[a]
	Regional total		7 907 848	9 710 713	12 826 116	16 880 592
San Joaquin Valley	Merced	North	215 256	319 785	537 166	792 667
	San Joaquin	North	579 172	884 375	1 454 089	2 122 660
	Stanislaus	North	459 025	708 950	1 160 376	1 690 026
	Sub-regional total		1 253 453	1 913 110	3 151 631	4 605 353
	Fresno	South	811 179	1 114 403	1 753 356	2 503 297
	Kern	South	677 372	1 073 748	1 919 849	2 923 829
	Kings	South	126 672	186 611	309 815	454 484
	Madera	South	126 394	224 567	411 713	635 019
	Tulare	South	379 944	569 896	982 425	1 468 811
	Sub-regional total		2 121 561	3 169 225	5 377 159	7 985 439
	Regional total		3 375 014	5 082 335	8 528 790	12 590 792
Sacramento	Sacramento	Central	1 212 527	1 651 765	2 409 784	3 312 096
	El Dorado	Foothills	163 197	256 119	381 668	530 209
	Nevada	Foothills	97 020	136 405	185 998	247 103
	Placer	Foothills	243 646	391 245	598 462	842 385
	Sub-regional total		503 863	783 769	1 166 127	1 619 697
	Sutter	North	82 040	116 408	173 672	241 405
	Yuba	North	63 983	84 610	124 998	172 890
	Sub-regional total		146 023	201 018	298 670	414 295
	Yolo	West	164 010	225 321	341 228	477 893
	Regional total		2 026 423	2 861 873	4 215 809	5 823 981
Non-metropolitan counties	Alpine		1 239	1 701	2 261	2 965
	Amador		34 853	40 129	46 935	57 739
	Butte		207 158	307 296	483 980	691 341
	Calaveras		42 041	62 688	91 124	125 014
	Colusa		20 973	41 398	82 055	131 662
	Del Norte		31 155	41 898	56 955	75 549
	Glenn		29 298	49 113	88 790	135 982
	Humboldt		128 419	141 092	158 279	190 693
	Inyo		18 437	20 694	27 538	36 140

Table 3.1 (continued)

Major region	County	Sub-region	Population estimates and forecasts			
			2000 (source: DOF)	2020 (source: DOF)	2050[a]	2100[a]
	Lake		60072	93058	148122	212717
	Lassen		35959	49322	69607	94087
	Mariposa		16762	23390	32101	42785
	Mendocino		90442	118804	169149	229650
	Modoc		10481	12396	16629	21911
	Mono		10891	14166	19434	25897
	Plumas		20852	23077	26612	32507
	Shasta		175777	240975	329849	439059
	Sierra		3457	3575	3678	4245
	Siskiyou		45194	53676	68588	88199
	Tehama		56666	83996	131321	186892
	Trinity		13490	15594	18300	22549
	Tuolumne		56125	77350	106662	142505
	Regional total		1109741	1515388	2177969	2990087
California			34653395	45448627	66763758	92081030

Note: a. See Table B.2 in Attachment B in http://www.energy.ca.gov/reports/ 2003-10-31_500-03-058CF_A03.PDF.

3.2.5 Infill Shares and Growth Densities: Both Will Increase

Two opposing forces shape the location and density of new urban development in California. Development has usually been attracted to California's coastal areas for reasons of both economics – that's where the ports are; and amenities – the climate along the coast is more moderate.[8] Accordingly, housing and land prices in California have long formed a downward-sloping gradient eastward from the coastal centers of Los Angeles, San Francisco, and San Diego.

As coastal locations have been increasingly built out, developers have moved farther inland. In addition to being less expensive, inland locations have usually been less subject to land use and environmental regulation than their coastal counterparts, making development cheaper and easier.

California was built by developers, and developers are nothing if not opportunistic. Even as they continue their inexorable push inland,

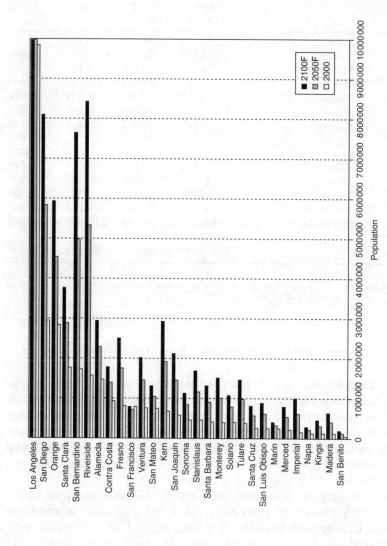

Figure 3.2 Projected population among metropolitan California counties, 2020, 2050, 2100

California's developers also continually look over their shoulders to consider potential infill and redevelopment opportunities. Thus, at the same time that California's coastal metropolitan areas are growing eastward, they are also infilling and redeveloping. And to the extent that infill development tends to occur at higher-than-existing densities, overall urban densities also rise.

This is the theory; in practice, local land use controls and opposition from neighborhood groups often function to make infill and redevelopment proportionately more difficult than greenfield development, breaking the link between growth at the urban fringe, increased infill activity, and rising urban densities. The result is less urban redevelopment and more urban sprawl.

In examining the share of each county's land area that was urbanized in 1972 with the population density of subsequent new development, we find that as predicted, marginal densities – measured as the change in population divided by change in urban land area – rise with the share of each county's land in urban use. Based on the fitted trend line, for every percent share of each county's land area in urban use in 1972, marginal development densities during the 1972–96 period rose by 26 individuals per acre.

Furthermore, when we examine the share of each county's land area that was urbanized as of 1972 with the share of new development occurring within the existing urban footprint in the form of infill, we find that county infill shares rise (and greenfield shares fall) with the share of each county's land in urban use. Based on the fitted trend line, for every percent share of each county's land area in urban use in 1972, the share of subsequent urban development occurring as infill – that is, within the initial 1980 urban footprint – rose by 200 percent.

Neither the density trend nor the infill trend fit the observed data all that well, a fact confirmed by the middling goodness-of-fit statistics of the estimated regression lines (see Table 3.2). Some counties such as Los Angeles, Orange, Santa Clara, San Mateo, and Stanislaus developed at higher densities and with more infill than average for their respective regions. Others, most notably Alameda, Contra Costa, and Sacramento, developed at lower densities or with less infill than expected.

Used with care, these two regression lines can be used to project future development densities and infill shares. In both cases, this involves incorporating additional information:

1. We project incremental densities by selecting the maximum of the recent incremental density for each county (denoted by the subscript i below) and the regression-based incremental density. This adjustment

*Table 3.2   Regression results comparing county 1972 urban land shares
with 1972–96 infill rates and development densities*

	Dependent variable: 1972–96 development densities		Dependent variable: 1972–96 infill rates	
	Coefficient	t-statistic	Coefficient	t-statistic
1972 urban land share	26.29	4.27	1.94	4.88
Intercept	6.21	11.04	0.13	3.55
Adjusted r-squared	0.34		0.40	
Number of observations	35		235	

has the effect of preventing projected incremental densities from falling.[9]

$$\text{Projected incremental density}_i = MAX\,[\text{recent incremental density}_i, \text{regression-based incremental density}_i] \qquad (3.5)$$

2. County-level infill growth shares are projected as the average of the current infill share and the maximum of the current infill share and the regression-estimated infill share. This adjustment has the effect of preventing infill shares from either rising too quickly or falling.
3. Projected greenfield population growth – the amount of population growth not projected to take the form of infill development – is calculated by multiplying projected population growth for each county by 1 minus the projected infill share for that county. The result of this calculation is then multiplied by the projected incremental density to yield an estimate of the amount of additional projected greenfield development.

$$\text{Projected greenfield population growth}_i = \text{Population projection}_i^* \, [1 - \text{projected infill share}_i]$$

$$\text{Projected greenfield development in acres}_i = \text{Projected greenfield population growth}_i {}^* \, [\text{Projected incremental density}_i] \,/\, 2.47 \qquad (3.6)$$

3.2.6   Updating the Inputs

Projected population growth is allocated to sites during three periods: 1997–2020, 2020–50, and 2050–2100. Several parameters and data layers are updated before each successive allocation round. These include the following:

1. Job accessibility. A job accessibility measure is calculated for each site based on its proximity via the highway network to all jobs located at discrete job centers. Subsequent to each growth allocation, a new set of job accessibility measures is computed for each site based on projected job growth by city or place and any changes in relative highway accessibility. For the baseline scenario (see section 3.3), no changes in relative highway accessibility are assumed.
2. City boundaries. Because development in California generally favors locations within cities – with some important differences among regions – it is essential to update city boundaries subsequent to each growth allocation, and to then estimate development probabilities accordingly. This does not present a problem for newly developed sites within existing city boundaries, but for sites outside existing boundaries, those boundaries must be stretched to accommodate the additional growth, which is done manually.
3. Physical features. The physical features of sites, such as their slope, location in a flood zone, or status as prime farmland, do not change between allocation rounds.
4. Urban share. After each allocation round, the share of land area in each county in urban use is updated. The updated urban share is then used to estimate updated incremental development densities and infill shares for the next allocation round.

3.2.7 Key Assumptions and Caveats

Numerous assumptions are embedded in this procedure and its components. Perhaps the most questionable is whether current population and employment growth trends and urban settlement patterns can be accurately extrapolated far into the future – in this case, through 2100 – and particularly in a state as changeable as California. If history teaches us anything, it is that the future is always different than we anticipate, no matter how sophisticated our reasoning or projection techniques. For this reason, the projections developed here and in later efforts are best viewed not as forecasts per se, but as scenarios – that is, as a set of illustrative futures designed to indicate how particular growth trends and development dynamics might play out on California's diverse landscapes. Beyond this general caveat, five specific assumptions drive this analysis:

1. The same factors that shaped land development patterns in the recent past will continue to do so in the future, and in the same ways. As we discussed previously, this procedure allocates future development to individual sites based on their projected development probability.

These probabilities are estimated using the results of a statistical model calibrated for 1988–98. The exact role of particular factors varies by region, but several influences are consistently important. These include proximity to freeways, access to jobs, site slope, and site incorporation status. Other factors such as farmland and wetland status vary more widely in their importance.

2. Jobs will continue decentralizing within California's four major urban regions – southern California, the greater San Francisco Bay area, the Sacramento region, and the southern San Joaquin Valley. Taking advantage of improved freeway access, less expensive land, and lower development costs, job growth during the last 50 years has favored suburban locations over core cities. To the extent that this trend continues – given the increasing importance of telecommunications in shaping economic geography, and in the absence of countervailing policies, there is no reason to believe that it should not – decentralizing job growth will continue to pull population outward, leading to more decentralized growth patterns.

3. California's population will continue to grow, and at more or less the same rate and in the same spatial pattern as projected by the California DOF. For consistency's sake, we rely on county population projections developed by the DOF through 2040. Thereafter, we extrapolate and trend downward the annualized county growth rates embedded in the DOF population projections. This approach yields a statewide population of 68 million in 2050 and 92 million in 2100.

    As large as these numbers are, they are hardly inconceivable. Since 1940, because of its robust economy, benign weather, and location on the Pacific Rim, California has been adding population at a steady rate of about 5 million individuals every decade. Should this trend continue, California's 2100 population would exceed 85 million. On the flip side, if California's economy falters, or if the state's high cost of living starts to choke off further job growth, the state's population could easily plateau at around 50 million and, although it seems unthinkable today, perhaps even trend downward.

4. Average infill rates and population densities will increase with additional development. As available supplies of developable land are used up, developers seek ways to use remaining land more intensely, either by increasing densities or through redevelopment. Thus, both development densities and infill activity should increase with population growth. Counteracting this tendency is the desire of many residents to preserve a rural or suburban lifestyle. In many parts of California, then, infill activities and development densities are below the levels suggested by theory. For the purposes of constructing a baseline scenario,

we assume that future infill activity and development densities will follow the upward trend lines from the last decade.

5. With respect to the baseline scenario (see next section), no new freeways or intra- and interregional rapid transit systems will be developed. Freeway road travel speeds will remain at current levels. This is perhaps the least realistic assumption of all. It is abundantly clear that California's growing population will need additional transportation infrastructure, but it is unclear what the infrastructure should be, where it should go, and how it should be planned and financed. Lacking these specifics, and for the purposes of constructing a baseline scenario, we assumed no change in transportation technology or facilities beyond what is currently available. The effect of this assumption is to direct additional growth to locations already served by transportation infrastructure rather than to new or different areas.

3.3   THE BASELINE SCENARIO

3.3.1   Building the Baseline Scenario

The function of the baseline scenario is to serve as a minimum-change alternative against which future scenarios that posit more extensive policy, regulatory, or investment interventions can be compared. More succinctly, the baseline scenario assumes continued growth along the lines of past trends and patterns without significant policy change. On the list of possible policy interventions not envisioned in the baseline scenario are additional infrastructure projects, environmental restrictions on land development, conservation and land preservation initiatives, and locally initiated changes in development densities and infill activities.

Scenario building involves four steps. First, we calculate a future development probability for each undeveloped site, using the land use change model results and job projections. For purposes of calculating future job and highway accessibility, no additions to the current highway system were assumed. Second, we specify a population growth increment to be allocated and an appropriate allocation density.

Third, we specify a list of absolute exclusion conditions denoting which sites may not be developed regardless of their development probability or the level of projected population growth. Four types of sites were excluded from development under the baseline scenario: (1) sites that are publicly owned; (2) sites that are currently under water; (3) sites that are identified as wetlands; and (4) sites with an average slope of more than 20 percent. Sites on which development is allowed under the baseline scenario include flood zone sites, farmlands of all types, sites in riparian areas, and sites presumed to be habitat to one or more threatened or endangered species.

Finally, we allocate prospective population growth to non-excluded sites in order of their development probability.

3.3.2   Baseline Scenario Results

Statewide baseline scenario results for 1998, 2020, 2050, and 2100 are presented in tabular form by region and county in map form in Figures 3.3 through 3.6 and in Table 3.3. Throughout the state, projected urban development will occur on flat sites, follow freeways, and be located in and adjacent to existing cities and urban places (Figures 3.3 through 3.6). Beyond these commonalities, growth patterns will differ significantly by region and county.

Southern California
Urban growth in the San Diego–Orange–Imperial subregion would account for about one-quarter of all new urban development in southern California. These trends are projected to continue into the foreseeable future. By 2050, Camp Pendleton, which separates San Diego and Orange counties, would be completely encircled by urban development. By 2100, if current trends continue, northern San Diego County and southwestern Riverside County will be completely urbanized, and intense urban growth would have moved east along I-10 into central San Diego County.

Most of Orange County's projected population growth is estimated to take the form of high-density infill. By 2050, almost all undeveloped lands in Orange County west and north of the foothills would have been developed. All of Imperial County's projected urban growth between 1998 and 2100 will occur along I-8, most of it within 10 miles of El Centro. Except for a few areas in the San Fernando Valley, Los Angeles County is almost entirely built out southwest of the San Gabriel mountains. As a result, most of Los Angeles County's projected population growth during the twenty-first century will take the form of infill and redevelopment.

Being closer to Los Angeles, Ventura County is estimated to grow more and sooner. Spatially, Ventura County would continue growing in a northwestern direction. Santa Barbara County should be able to continue resisting southern California's extreme growth pressures for about another 20 years. After 2020, development activity should pick up. With growth in southeastern Santa Barbara County limited by the Santa Ynez mountains, most new development will happen along the Highway 101 corridor from Buellton north to Santa Maria. Indeed, by 2100, the entire Highway 101 corridor from downtown Los Angeles north to Santa Maria will be essentially built out.

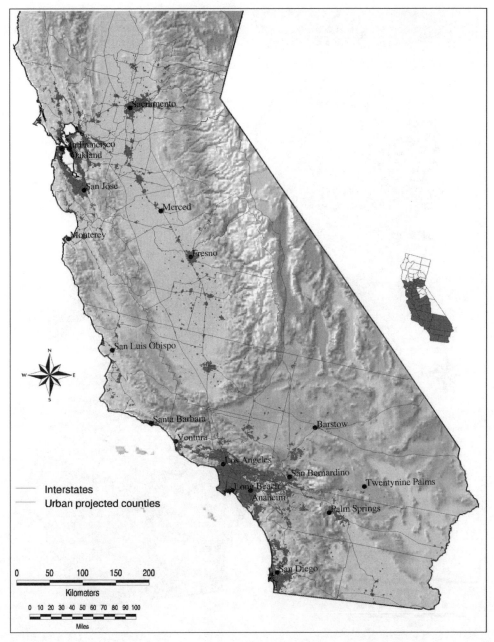

Figure 3.3   California's urban footprint, 1998 (population: 33 million)

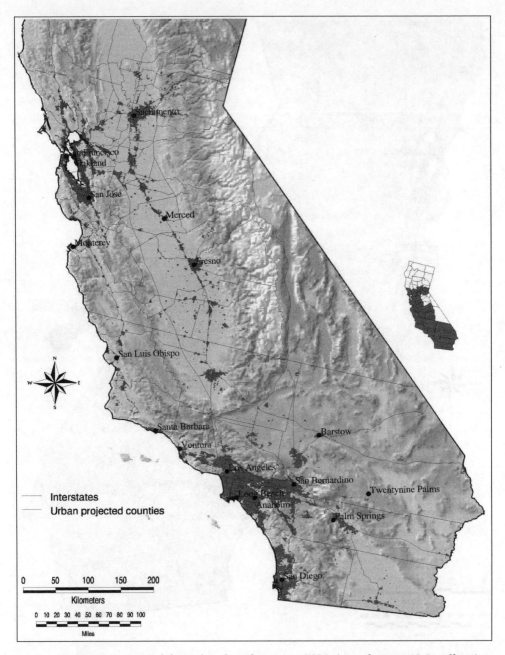

Figure 3.4   California's urban footprint, 2020 (population: 45.5 million)

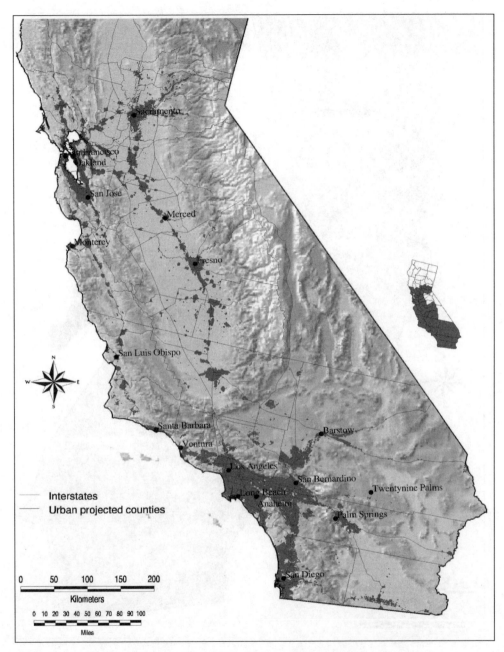

Figure 3.5   California's urban footprint, 2050 (population: 67 million)

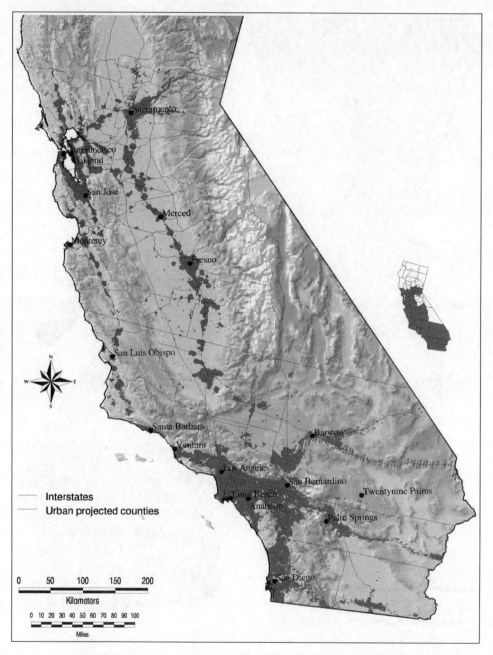

Figure 3.6   California's urban footprint, 2100 (population: 92 million)

Table 3.3   Urbanized land areas by county

Major region	County	Subregion	Urbanized land area (ha)			
			1998	2020	2050	2100
Southern California	Los Angeles	Central	307 205	318 174	342 037	360 808
	Imperial	South	9682	19 834	38 365	59 615
	Orange	South	109 364	116 424	122 459	129 443
	San Diego	South	125 883	164 271	224 118	290 171
	Subregional total		244 929	300 529	384 942	479 229
	Riverside	Inland Empire	97 760	162 938	270 893	389 620
	San Bernardino	Inland Empire	110 329	171 155	281 363	411 287
	Subregional total		208 089	334 093	552 256	800 907
	Santa Barbara	North	24 061	28 142	45 317	63 227
	Ventura	North	39 135	50 043	67 330	85 631
	Subregional total		63 196	78 185	112 647	148 858
	Regional total		823 419	1 030 981	1 391 882	1 789 802
Northern California	Alameda	Central	56 562	63 453	70 471	79 053
	Contra Costa	Central	55 547	60 250	65 067	70 751
	San Francisco	Central	9386	9386	9386	9386
	San Mateo	Central	28 473	29 769	31 682	34 300
	Santa Clara	Central	72 717	77 510	83 628	91 392
	Subregional total		222 685	240 368	260 234	284 882
	Marin	North	16 073	16 590	17 718	19 373
	Napa	North	8313	9861	11 924	14 411
	Solano	North	21 470	27 815	35 417	44 275
	Sonoma	North	26 762	34 494	43 614	54 514
	Sub-regional total		72 618	88 760	108 673	132 573
	Monterey	South	20 224	28 922	50 837	74 896
	San Benito	South	2709	4344	7240	10 457
	San Luis Obispo	South	14 989	20 920	32 512	45 581
	Santa Cruz	South	11 539	14 713	21 142	21 145
	Subregional total		49 461	68 899	111 731	152 079
	Regional total		344 764	398 027	480 638	569 534
San Joaquin Valley	Merced	North	12 358	18 528	29 353	41 382
	San Joaquin	North	29 023	43 284	63 652	85 114
	Stanislaus	North	20 430	25 142	38 362	54 256

Table 3.3   (continued)

Major region	County	Subregion	Urbanized land area (ha)			
			1998	2020	2050	2100
	Subregional total		61811	86954	131367	180752
	Fresno	South	37765	48893	81243	119323
	Kern	South	40840	65117	111187	159400
	Kings	South	11501	15094	20830	27189
	Madera	South	9025	15348	24970	35827
	Tulare	South	19701	30181	52627	76216
	Subregional total		118832	174633	290857	417955
	Regional total		180643	261587	422224	598707
Sacramento	Sacramento	Central	61009	71950	85317	100003
	El Dorado	Foothills	10436	13920	17829	22611
	Nevada	Foothills	5924	7935	9905	12367
	Placer	Foothills	15284	23776	33099	44089
	Subregional total		31644	45631	60833	79067
	Sutter	North	4311	6385	9202	12333
	Yuba	North	4531	5821	7834	10290
	Subregional total		8842	12206	17036	22623
	Yolo	West	10368	12923	16752	21422
	Regional total		111863	142710	179938	223115

With coastal areas running out of buildable land, the real development action in southern California during the twenty-first century is projected to be in Riverside and San Bernardino counties – known as the Inland Empire. Urban development in the Inland Empire would increase nearly 400 percent. Sixty percent of new urban development in southern California during this century will be in the Inland Empire. San Bernardino and Riverside counties' growth rates and patterns would also be similar: development will proceed from west to east, along I-10 and I-215 in Riverside County, and along I-10, I-15, and I-40 in San Bernardino County. By 2050, the Victorville–Apple Valley–Hesperia area of San Bernardino County and the Perris–Hemet–Moreno Valley area of Riverside County are estimated to emerge as major metropolitan centers, and the Inland Empire will be entirely built out west of the line connecting Hemet in Riverside County and Yucaipa in San Bernardino County. By 2050, the Coachella Valley (stretching from Palm Springs to Indio) would be built out south of I-10; by 2100, the north side of the Coachella Valley would have been developed.

The San Joaquin Valley
Nearly three-quarters of new urban development in the entire San Joaquin Valley (stretching from Kern County in the south to San Joaquin County in the north) is projected to be in the southern subregion. As it has since the turn of the twentieth century, new development in the southern San Joaquin Valley would be concentrated along the Highway 99 corridor – with or without the construction of a high-speed rail system. By 2100, the entire corridor would be urbanized, and active farmlands would have been pushed to the east and west.

About half of the northern subregion's urban growth would occur in San Joaquin County. By 2100, San Joaquin County's urban footprint will rival Santa Clara's in size. Stanislaus County would also grow substantially, with most of its new development in and around Modesto and Turlock. Growth in Merced County will generally proceed north to south: starting in Livingston, then moving south to Merced.

Almost one-third of the region's urban growth would be in Kern County. Bakersfield would continue to dominate Kern County's urban landscape. Even so, new and smaller urban nodes would also develop around the cities of Shafter and Delano by 2020, Wasco and Tehachapi by 2050, and Arvin and Mojave by 2100. Almost all new urban development in Fresno County is estimated to occur at the outskirts of the city of Fresno or along Highway 99, and the entire Highway 99 corridor in Tulare County would be urbanized.

New development in the northern San Joaquin Valley subregion would be principally fed by unaccommodated eastward overflow growth from the Bay Area. Altogether, urban development in the northern San Joaquin Valley is projected nearly to triple. As in the southern part of the San Joaquin Valley, new development in the north would be concentrated along the Highway 99 corridor. By 2100, the entire Highway 99 corridor would be developed to a width of 10–20 miles in San Joaquin County, down to 5–10 miles in Merced County.

About half of the northern subregion's urban growth would occur in San Joaquin County. By 2100, San Joaquin County's urban footprint will rival Santa Clara's in size. Stanislaus County would also grow substantially, with most of its new development in and around Modesto and Turlock. Growth in Merced County will generally proceed north to south: starting in Livingston, then moving south to Merced.

Northern California
We included the Monterey Bay Area and San Luis Obispo County as a subregion of northern California, which is usually not done – at least not yet. More and more, however, as the Silicon Valley continues to grow southward, its economic sphere of influence will envelop Monterey, San Benito, Santa Cruz, and even the northern section of San Luis Obispo County. Depending on the time period, urban growth in this subregion will

account for between one-third and one-half of new urban development in northern California during the twenty-first century. Unlike the central Bay Area, where significant population growth would occur as infill, most population growth in the Monterey Bay Area and points south would occur as greenfield development.

Monterey County, being more directly connected to Santa Clara County along Highway 101, is projected to grow more and sooner than its three subregional neighbors. Indeed, by 2050, the central spine of the Salinas River Valley – which includes some of the most fertile farmland in the world – will essentially be built out. Farther to the south, growth in San Luis Obispo County's would occur radially around the cities that line the Highway 1010 corridor.

With so little undeveloped land left adjacent to the San Francisco Bay, most new development in this subregion – encompassing Alameda, Contra Costa, San Francisco, San Mateo, and Santa Clara counties – would occur east of Oakland and the East Bay Hills and south of San Jose. Most of the subregion's population growth would take the form of infill and redevelopment.

Already mostly urbanized, Santa Clara County has little flat and accessible land available for future development. Almost all of this increase will occur within the Highway 101 corridor south of San Jose. Most of the increase in Alameda County's urban footprint would occur in and around three cities: Dublin, Pleasanton, and Livermore. Further north, Contra Costa County's new urban growth would be divided between the I-680 corridor that connects Martinez and Pleasanton and the Highway 4 corridor that connects Concord and Brentwood. Development will also be continuously climbing the foothills of Mount Diablo and the western side of the East Bay Hills.

On the San Mateo Peninsula, San Francisco is already entirely built out and would accommodate all its projected population growth through infill and redevelopment. San Mateo County would also grow mainly through infill and redevelopment. Most of this growth will occur adjacent to the San Francisco Bay or south of Pacifica along Highway 1.

The north San Francisco Bay Area includes Marin, Sonoma, Napa, and Solano counties. Most of the increase in urbanization would take place in Sonoma and Solano counties. By 2050, the Highway 101 corridor would be continuously developed from the Sonoma–Marin county line north through the city of Healdsburg. Solano County's urban footprint would double. Compared to Sonoma and Solano counties, Marin and Napa counties would hardly grow at all. Lacking good freeway access, Napa County would also experience only moderate growth.

The Sacramento region

By 2100, if current trends continue, the region's urban footprint will have more than doubled. Located at the confluence of the Sacramento and American rivers, Sacramento County still has ample flat land on which to grow – mostly to the south and east. New growth would also extend north along I-5 and California Highway 70. All told, Sacramento County is projected to account for about one-third of the Sacramento region's growth during the twenty-first century.

The foothill subregion consists of the western sides of El Dorado, Placer, and Nevada counties. By 2050, urbanization in the foothill subregion will have nearly doubled; by 2100, it would increase another 60 percent. Most of the subregion's growth will occur in Placer County along I-80.[10] Indeed, by 2100, the I-80 corridor would be completely built out past Auburn, to Meadow Vista. Lesser (although still sizable) amounts of development are projected for El Dorado and Nevada counties. Growth in El Dorado County would be focused along Highway 50, in and around the Placerville area. In Nevada County, new urban development would favor the Grass Valley–Nevada City area. All together, the three-county foothills subregion would account for about 40 percent of the growth of the Sacramento region during the twenty-first century.

Yolo County's urban footprint is projected to double. Unless actions to limit growth are taken, which the city of Davis does periodically, just about all of this increase would occur in and around the three cities that line I-80: Dixon, Davis, and West Sacramento.

If current trends continue, Yuba and Sutter's urban footprint is also estimated to double by 2050 and to increase another 35 percent by 2100. These are big increases by the standards of Yuba and Sutter counties. Spatially, most of this subregion's growth would occur in southeastern Sutter County, near the Sacramento County border.

3.4   BASELINE IMPACT ASSESSMENT

The conversion of undeveloped land to urban uses generates three types of effects on the landscape: (1) it reduces the amount of undeveloped land still available; (2) it alters the patch size, shape, and fragmentation level of the remaining undeveloped landscape; and (3) it alters both the amount and quality of the resource and environmental services provided by undeveloped lands. This section assesses the effects of (1) under the baseline scenario for 1997–2020, 2020–50, and 2050–2100. These effects can be measured using many of the same digital data layers used to derive the baseline scenario. Considering the effects of urban growth on the supply and quality of resource and ecological services is beyond the scope of this effort.

3.4.1 Hillsides and Steeply Sloped Land

Statewide, projected urban growth presents a relatively small threat to steep hillsides. Among the 45 counties for which we developed detailed urban growth projections, we project that an additional 8200 ha of steeply sloped land – that is, land with a slope in excess of 15 percent – is estimated to be developed by 2020. By 2050 and 2100, we project that urban growth will have consumed an additional 38 000 and 55 000 ha of steeply sloped land, respectively. These growth increments account for only 0.1 percent, 0.4 percent, and 0.6 percent of the current hillside land area of these 45 counties.

The counties projected to suffer the largest absolute hillside losses by 2050 and 2100 are all in southern California: San Diego County (–14 600 ha, or 4 percent of the county's remaining steep hillsides), Riverside County (–13 900 ha, or 3 percent), Los Angeles County (–8800 ha, or 3 percent), and San Bernardino County (–6300 ha, or 1 percent). Placer and El Dorado are the only counties outside the southern California region that are projected to suffer significant hillside losses because of urbanization. Because it is extremely flat, Sacramento County is likely to suffer minimal absolute hillside losses, but significant relative losses.

Note that these projections assume a continuation of current development trends and patterns. If these patterns shift in ways that make hillside development easier from a regulatory perspective, less costly from a development perspective, or more attractive from a market perspective, it is quite conceivable that amounts of hillside loss could be much greater, particularly in counties like San Diego, Ventura, Orange, Santa Barbara, Santa Clara, and Marin. All these counties are running out of accessible flat land near urban centers.

3.4.2 Wetlands

Mainly for planning and regulatory reasons, wetland development is growing increasingly difficult throughout California. Counties with large amounts of wetlands that are in agricultural use are looking for ways to keep them that way, and counties with few remaining wetland areas are vigorously trying to protect and enhance them.

Statewide, projected urban growth presents a small but significant threat to wetlands, particularly those identified as part of the National Wetland Inventory (NWI).[11] In the 28 California counties with significant remaining wetlands that are threatened by urban growth, we project that an additional 12 000 ha of wetlands will be developed by 2020. By 2050 and 2100, respectively, we project that urban growth will consume an additional 26 000 and 42 000 ha of wetlands. In percentage terms, these growth increments account for only 1 percent, 2 percent, and 3 percent, respectively, of the current wetlands inventory.

The counties projected to suffer the largest absolute wetlands losses by 2050 and 2100 are mostly in the northern San Joaquin Valley, in the southern Sacramento River Valley, or adjacent to the San Francisco Bay. They include San Joaquin County (–8600 ha, or 11 percent of the county's remaining wetlands), Sutter County (–4300 ha, or 7 percent), Sonoma County (–3200 ha, or 26 percent), Solano County (–2500 ha, or 5 percent), and Alameda County (–2300 ha, or 33 percent). A number of additional counties are facing moderate absolute wetland losses but large percentage losses: Marin County (–1700 ha, or 24 percent), San Mateo County (–1,300 ha, or 42 percent), San Diego County (–1200 ha, or 15 percent), and Santa Clara (–700 ha, or 15 percent). At the other extreme, Sacramento, Merced, and Yolo counties are all facing moderate absolute wetland losses but small relative losses.

3.4.3 Riparian Areas

Riparian zones are the land areas around rivers, streams, lakes, and permanent wetlands. They are typically but not exclusively characterized by woody, fast-growing vegetation and water-oriented bird, animal, and insect species. Inventories of riparian areas have thus far been developed for the San Francisco Bay and San Joaquin Valley but not for the rest of the state. To augment these more limited data sources, we generated a statewide, 100 m riparian zone data layer by buffering all inland rivers, streams, and lakes listed in the 2000 Census TIGER file (US Census Bureau, 2001; Digital TIGER line files). Although it is comprehensive and consistent, this method tends to overestimate the total amount of riparian area and underestimate the area of specific riparian zones.

The counties projected to suffer the largest absolute riparian losses by 2050 and 2100 are mostly in southern and coastal California. They include San Bernardino County (− 52 200 ha, or 6 percent of the county's remaining riparian zone land area), Riverside County (− 51 000 ha or 13 percent), San Diego County (–22 000 ha, or 11 percent), Imperial County (–14 000 ha, or 6 percent), and Kern County (− 14 000 ha, or 3 percent). A number of additional counties are facing moderate to small absolute riparian zone losses but large percentage losses by 2100: Stanislaus County (− 5000 ha, or 12 percent), San Joaquin County (–4900 ha or 12 percent), Sacramento

County (4600 hectares, or 12 percent), Alameda County (3800 hectares, or 13 percent), Orange County (–2700 ha, or 11 percent), Santa Cruz County (–2400 ha, or 10 percent), San Diego County (–1200 ha, or 15 percent), and Santa Clara (–700 ha, or 15 percent).

3.4.4   Farmlands

This section briefly discusses how certain important types of farmlands are predicted to be reduced by urbanization. The CFMMP defines prime farmland as land used for the production of irrigated crops at some time during the previous four years. Projected urban growth presents a significant threat to the state's remaining supplies of prime farmland, especially in the San Joaquin, Monterey, and Imperial valleys. In 2020, 2050, and 2100, respectively, we project that urban growth will consume 3 percent, 9 percent, and 17 percent, respectively, of the current inventory of prime farmlands. Most at risk from urban growth are prime farmlands in the San Joaquin and Monterey valleys. Five counties will lose more than half of the little prime farmland they still have: San Bernardino (78 percent), San Diego (58 percent), Orange (57 percent), Alameda (51 percent), and Santa Clara (50 percent).

The situation might not be quite as bleak as these numbers would make it seem. If current trends continue, even as large amounts of prime farmland are lost to urban growth, farmers will likely be 'developing' new prime farmlands in other locations, mostly by extending irrigation to grazing and secondary farmlands. Although not of the soil quality of the prime farmlands being lost, assuming sufficient water is available at the right price, these new farmlands should easily sustain California's agricultural economy. Potential opportunities for new prime farmland development are most plentiful in the San Joaquin and Monterey valleys (where urbanization will pose less of a threat to grazing lands), but are extremely limited in southern California and the Bay Area.

'Farmlands of state importance' are those similar to prime farmland but with minor shortcomings, such as greater slopes or less ability to store soil moisture. 'Farmland of local importance' is determined by each county's board of supervisors and a local advisory committee.

Projected urban growth presents a significant threat to the state's remaining supplies of state and locally important (S&LI) farmlands, especially in Riverside, Imperial, San Diego, Sacramento, and San Joaquin counties. Among the 36 California counties in which S&LI farmlands are threatened by urban growth, we project that growth increments account for only 4 percent (2020), 9 percent (2050), and 14 percent (2100), respectively, of the current inventory of S&LI farmlands.

Most at risk from urban growth are S&LI farmlands in the Inland Empire and in the central Sacramento–San Joaquin Valley. Even though Bay Area and coastal counties will suffer significant relative losses, except for Monterey and Sonoma counties, their absolute losses will not be that great. Whether or not these losses will be offset through the irrigation and conversion of grazing land will depend on many factors: water availability, commodity prices, labor costs, and changing environmental regulations.

Projected urban growth presents a significant threat to the state's remaining supply of unique farmland that supports the state's leading agricultural crops, especially in San Diego and Riverside counties. We project that urban growth will remove 2 percent, 5 percent, and 8 percent, respectively, of the current inventory of productive farmlands in the counties studied.

Most at risk from urban growth are productive farmlands in and around the Inland Empire and in the central San Joaquin Valley. Three counties in addition to San Diego and Riverside will lose more than half of their productive farmlands to urban development by 2100: San Bernardino, Alameda, and Orange. All of California's coastal counties south of San Francisco and most of its Central Valley counties are facing significant losses of unique farmlands as a result of projected urban growth.

The CFMMP identifies 'grazing lands' as those on which the existing vegetation is suited to the grazing of livestock. Projected urban growth presents a significant threat to grazing lands in Riverside, Placer, San Diego, and San Bernardino counties; a moderate threat in Orange, Ventura, Alameda, Solano, Sacramento, Los Angeles, Santa Cruz, and Santa Barbara counties; and a minor threat elsewhere in the state. Among the 35 California counties in which grazing lands are threatened by urban growth, we project that only 1 percent (2020), 2 percent (2050), and 3 percent (2100), of the current inventory of grazing land in the counties studied would be consumed by urbanization.

Most at risk from urban growth are grazing lands in and around the Inland Empire. The real threat to grazing lands may be bigger than these numbers suggest. As urbanization consumes prime and unique quality farmlands in the San Joaquin, Monterey, and Imperial valleys, agricultural businesses may attempt to convert grazing lands to cultivated use, mostly by extending irrigation. The realization of this potential 'domino effect' will depend on many factors, including land, water, and agricultural product prices, and the costs of extending key infrastructure.

The trend toward ranchette and resort and vacation home development also threatens California's grazing lands. To the extent that California's urban areas continue to spin off ranchettes and rural subdivisions, the potential threat of population growth to grazing land may be the greatest of all. As a result, although the geography of cultivated farmland in California in 2100 could very well be similar to that of today, the geography of California's grazing lands will almost certainly be different.

3.5   CONCLUSION

California's land use changed tremendously in the twentieth century and could continue to change dramatically in the twenty-first century. We examined two plausible scenarios of population growth for California: 67 million and 92 million total population by the end of the century. Currently heavily urbanized areas such as the Bay Area and Los Angeles would be, for the most part, built out. Development would stretch well into the Central Valley and southeastern California. Such population growth and the development it would spur would have substantial implications for urbanization, farmlands, natural areas, and use of natural resources such as water.

Subsequent chapters use these results to analyze implications for biodiversity (Chapter 6), water resources (Chapter 10), and agriculture (Chapter 11). Such socio-economic changes as increase in population can have impacts of a magnitude comparable to the impacts of climate change.

NOTES

1. More details are provided at http://www.energy.ca.gov/reports/2003-10-31_500-03-058CF_A03.PDF.
2. In areas where views are rewarded in the marketplace with price and rent premiums, the probability of development may actually rise with slope.
3. This approach requires defining regions in terms of commute sheds, the residential areas from which job centers draw daily workers.
4. By net job growth we mean the excess of job gains (through attraction and expansion) over losses.
5. DOF uses a modified cohort-component model, disaggregated by race and one-year age cohort to project population.
6. See Appendix III, Table B at http://www.energy.ca.gov/reports/2003-10-31_500-03-058CF_A03.PDF.
7. County population projections for 2100 using these reduced growth rates are presented in the final two columns of Table A.2 in Attachment A at http://www.energy.ca.gov/reports/2003-10-31_500-03-058CF_A03.PDF.
8. Note this analysis did not examine whether climate change would affect housing preferences.
9. See Tables B.1 and C.2 at http://www.energy.ca.gov/reports/2003-10-31_500-03-058CF_A03.PDF.
10. The projections detailed in Table 3.3 significantly undercount the number and area of ranchette developments.
11. The NWI does not include vernal pools.

REFERENCES

Landis, J., M. Reilly, P. Monzon, and C. Cogan. 1998. *Development and Pilot Application of the California Urban and Biodiversity Analysis (CURBA) Model.* University of California, Berkeley, CA.

McFadden, D. 1974. The measurement of urban travel demand. *Journal of Public Economics* 3(4): 303–28.

US Census Bureau. 2001. TIGER/Line Files, Census 2000. US Department of Commerce, US Census Bureau Geography Division, Washington, DC. Available at http://www.census.gov/geo/www/tiger/index.html.

4. Climate change scenarios

Chuck Hakkarinen and Joel B. Smith

This chapter describes the climate change scenarios used in this study. Before presenting possible future scenarios, however, we provide a brief background to California's current climate and recent trends. The selection of future scenarios for this study was based on a workshop (described in http://www.energy.ca.gov/reports/2003-10-31_500-03-058CF_A01.PDF) with climatologists in California and impacts researchers.

4.1 INTRODUCTION

It is not possible to predict exactly how California's climate will change during this century. Climate models are unable to make definitive statements about how regional climate will change, although their capability is improving (Kerr, 2001). The models yield a range of plausible climate changes that become more uncertain the further into the future the prediction. In spite of this uncertainty, it is very important to study how California may be affected by climate change. Such information may be important for the state as it examines options to control its greenhouse gas emissions, and the information will certainly be useful to the state when examining adaptations to reduce adverse impacts of climate change or to take advantage of positive impacts.

To understand how future climates might affect California, we examined a range of plausible scenarios of climate change. The range helps identify the sensitivities of affected systems and provides a sense of what kinds of impacts are possible under climate change. This kind of information can be useful to help understand how vulnerable California may be to climate change and in beginning to address policy issues such as adaptation.

The scenarios represent different possible outcomes, but they are not predictions of climate change. The scenarios have not been weighted by likelihood, and there are combinations of changes between precipitation and temperature that we did not examine. Also, seasonal and daily fluctuations could vary, and there could be a different distribution of changes across the state. The scenarios are not the only outcomes that could happen; the actual change in climate may be quite different from any of the scenarios.

4.2 CURRENT CONDITIONS

Although California has quite varied climate (from extreme heat in the low-lying desert to extreme cold in the Sierra Nevada), the state is generally known for its mild climate. This is the result of the moderating influence of the Pacific Ocean on those areas of the state where the majority of the populace lives. The Pacific Ocean currents keep the state, particularly the coastal areas, relatively cool in the summer and mild in the winter. The Sierra Nevada along the eastern edge of the state tend to block intrusions of extreme cold weather in the winter from most of the state. As a result of its complex topography, California can have large regional variations in daily weather; for example, temperatures can vary as much as 50°F within 50 miles on typical summer days (Western Regional Climate Center, 2005).

A key determinant of the seasonality of California's climate is the location of the subtropical high pressure system in the eastern Pacific Ocean and its migration north and south during seasons and between years. When the subtropical high moves farther south (in the winter and often in El Niño years), the resulting position of the jet stream enhances the probability of Pacific storms striking the California coastline. Migration of the subtropical high farther north (in the summer and often in La Niña years) tends to reduce the probability of winter storm landfall in California. The positions of the subtropical high appear to be strongly influenced by very large-scale climate factors that extend across much of the Pacific Basin. Global warming is unlikely to change this critical feature of California's climate.

Ocean circulation also plays a major role in influencing year-to-year variation in California's climate. Cool ocean water in the Pacific Current along the coastline suppresses both air temperatures and evaporation near the coast. Two key drivers of California's interannual climate variability are the El Niño/Southern Oscillation (ENSO) and the Pacific Decadal Oscillation (PDO). The El Niño condition arises when the eastern Pacific surface waters near South America warm. This condition, which tends to occur every three to five years and usually lasts about 12 to 18 months, can alter global weather patterns. Statistical studies of El Niño occurrences during the last century show that they often result in warmer temperatures in California (mostly in the winter) and a southerly shift of the extratropical jet stream. This often brings more winter precipitation into southern California than average. For example, the 1997–98 El Niño, which was one of the strongest on record, brought several extreme precipitation events into southern California. These events, combined with denuded hillsides from previous seasonal wildfires, resulted in significant flooding and mudslides. The warm and wet conditions often associated with El Niño may also increase the presence of mosquitoes and molds.[1]

La Niña, the 'opposite' climate state to El Niño, is characterized by cooler than normal sea surface temperatures in the eastern tropical Pacific Ocean. The winter Pacific storm tracks in La Niña years tend to be farther north, producing more precipitation in Oregon and Washington and less precipitation in California. However, the general tendencies in temperatures and precipitation associated with El Niño and La Niña do not always hold true. For example, some of the wettest winter seasons (total winter precipitation) in northern California have occurred in La Niña years. And although significant flooding has often occurred in southern California in El Niño years, no major flood was ever recorded in northern California in an El Niño year during the twentieth century.

In contrast to ENSO, the PDO fluctuates on a 20- to 30-year cycle, involving either warming (warm phase) or cooling (cold phase) of waters in the North Pacific Ocean. Warm PDOs have been correlated with increased fishery productivity off Alaska and decreased productivity off the west coast of the lower 48 states. Cold PDOs have yielded the opposite effect. A warm phase of the PDO is believed to have begun around 1977 and may have ended in the mid-1990s. Although some studies have correlated ENSO with climate change (for example Trenberth and Hoar, 1996), the PDO has not been correlated with climate change. Given the long periodicity of the PDO, few scientific observations of full PDO cycles have been possible to date, and its mechanisms of formation and evolution are poorly understood.

Changes in California's climate in the twentieth century were not consistent across the state. For example, while Redding cooled, Davis and Pasadena warmed. There was an apparent warming in the spring in northern areas, with as much as a 2°C (4°F) warming observed in some northern areas. Although Eureka, Sacramento, and San Diego experienced little change in precipitation, the northern Sierras saw an increase. One clear trend was a reduction in snow water content, particularly in April (see Figure 4.1).

4.3   SELECTION OF CLIMATE CHANGE SCENARIOS

Climate change scenarios should meet two criteria:

1. They should be physically consistent with scientific understanding about how global climate will change.
2. They should represent a broad, plausible range of change in climate so as to capture uncertainty about either direction (that is, increase or decrease in a variable such as temperature) or magnitude of change.

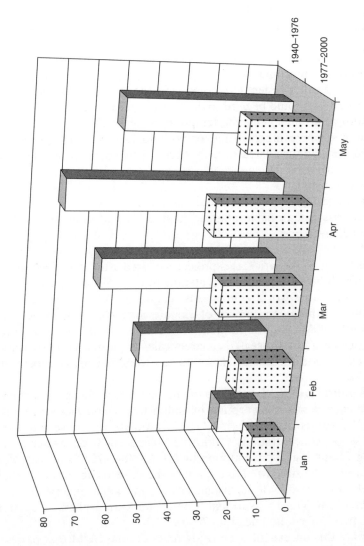

Figure 4.1 Monthly average snow water content (inches) in the Sierra Nevada Mountains measured at Echo Summit, California, 1940–2000

> Using a narrow range of scenarios may convey more certainty about
> the future than is scientifically justified.

In addition, it was important for this study that the scenarios cover the
entire period from 2000 to 2100.

The scenarios selected for the study meet the following criteria. They:

- Reflect a reasonable range of change in key parameters such as temp-
  erature and precipitation.
- Provide a full transient (that is, at least the 2020s, 2060s, and 2090s)
  of results, or could be used to develop a transient scenario.
- Capture current spatial and temporal variability.
- Address changes in spatial and temporal variability to a feasible extent.
- Can be made available quickly (that is, based on completed and
  readily available climate model runs or transparent enough to be
  created easily).
- Can be readily used in impact assessments.
- Comprise a reasonable number (that is, fit within budgets of the
  impact assessors).

To select a set of scenarios, this project analyzed estimates of change in
California's climate by general circulation models (GCMs). GCMs are
models of the entire world's climate. By changing radiant forcing, these
models can also provide estimates of climate change on a regional scale.
The scale of GCMs is quite limited (see below), but an examination of
many of them is useful to see if, on a large scale, they agree about the direction and magnitude of change in key climate variables such as temperature
and precipitation.

Table 4.1 displays GCM estimates of the percentage change of precipitation for each degree of warming in California. These ratios can vary
dramatically across models. Some GCMs estimate that with increased
greenhouse gas concentrations California's climate will get wetter, some
projecting a doubling of precipitation by 2100. A few GCMs estimate that
precipitation in the state will remain about the same or decrease. Overall,
the GCMs do not show a consistent change in annual precipitation in
California (Tom Wigley, National Center for Atmospheric Research
(NCAR), personal communication, 2000).

To capture this uncertainty, the study relies on one GCM that projects
California will become wetter and another that projects California will
become drier. Even though they may disagree on whether precipitation will
increase or decrease in California, most of the models agree on the seasonal
pattern of changes, estimating that winter precipitation will increase and

Table 4.1 *Average precipitation changes for California (percent change per 1°C global-mean warming)*

GCM	Annual	December–February	June–August
BMRC	−8.0	−6.5	−9.9
CCC	6.9	14.0	3.5
CSIR1	−0.7	−1.8	−1.0
CSIR2	2.6	5.3	−2.7
ECH1	9.8	8.4	2.8
ECH3	−3.2	9.9	−22.5
GFDL	0.0	1.8	−0.1
GISS	2.2	1.5	3.6
LLNL	0.0	1.5	−2.7
OSU	−1.3	0.6	−5.2
UIUC	2.3	0.3	34.7
UKHI	2.6	6.2	−5.2
UKLO	4.1	6.1	−0.2
UKTR	2.9	12.4	0.3
CCCTR	26.3	56.0	7.1
JAPAN	−7.7	−10.7	0.7
CSITR	−2.8	7.7	−10.0
ECH4	−3.1	8.7	−8.1
GFDTR	−0.1	−3.4	−4.6
HadCM2	13.8	23.1	7.8
NCAR	2.1	0.4	7.4
Overall mean	2.3	6.7	−0.2
Standard deviation	7.3	13.2	10.4

Notes:
Grid box central points (5° by 5° grid).
Latitude range is 32.5° to 42.5° N inclusive.
Longitude range is −122.5° to −117.5° E inclusive.

Source: Tom M.L. Wigley, National Center for Atmospheric Research, personal communication, 21 June 2000.

summer precipitation – which is already quite low across the state – will decrease.

The GCM runs selected are:

- The relatively wet and warm Hadley (HadCM2) scenario (Johns et al., 1997), which was also used in the US National Assessment (NAST, 2000).
- A relatively cooler and drier run from NCAR's Parallel Climate Model (PCM) (Dai et al., 2001).[2]

The two GCMs alone gave a wide divergence in scenarios; the Hadley model estimated about a 65 percent increase in winter precipitation by 2100 and the PCM model estimated about a 20 percent decrease. This is likely to lead to broad differences in estimated impacts of climate change. Plates 1 and 2 display the changes in temperature and precipitation from the present day through the end of the twenty-first century at the 10 km scale level for the HadCM2 and PCM GCMs, respectively, displaying wet season (November through March) and summer (June through August) months.

The GCM data are generated for large grid cells ($2.5 \times 3.75°$ for HadCM2 and $2.8 \times 2.8°$ for PCM). These data projections were downscaled to 10 km cells using the procedure described in Box 4.1. This downscaling had already been performed for the HadCM2 scenario as part of the US National Assessment (NAST, 2000). The PCM results had to be downscaled specifically for this project.

BOX 4.1   PROCEDURE FOR DEVELOPING GCM-BASED CLIMATE CHANGE SCENARIOS

1. The GCM scenarios were statistically interpolated (downscaled) to a 10 km (100 km^2) grid. Although the downscaling allows for more disaggregation of results, it does not completely capture how California's microclimates may change in the future (Daly et al., 1999).
2. The GCM output for the base period (1961–90) was subtracted from or divided into the GCM output from three periods in the transient run:

   - 2010–39
   - 2050–79
   - 2080–99

3. The changes in temperature were added to an observed data set of temperature at the 10 km level and the precipitation ratios were multiplied by observed precipitation. Using the full transient allows for change in interannual variability and seasonal climate. It does not allow for changes in daily climate or in the diurnal cycle. Note that some studies used average change in climate in the three time periods. These studies assumed that interannual climate would not change from the observed period.

Regional climate models (RCMs) are typically high spatial resolution climate models (for example 50 km grid boxes), which are often 'nested' within GCMs (that is, driven by boundary conditions from GCMs). With their higher spatial resolution than GCMs, some RCMs that have been applied to California can resolve such features as the coastal mountain ranges, the Central Valley, and the Sierra Nevada mountains, all of which have a significant influence on shaping California's climate. Because only limited results were available at the time of this study from RCMs (for example Fu et al., 2005), we decided not to use the RCMs. The RCMs we examined estimated climate change for only one decade, the 2040s. We considered such a limited time scale to be inadequate for our needs.

In addition to selecting the two GCM runs, we also examined a set of uniform incremental scenarios, which impose uniform changes in the historical climate data for all temporal and geographical scales. For example, the 3°C uniform warming scenario assumes that temperatures will rise in Los Angeles and in Sacramento by 3°C in both winter and summer. Three temperature change scenarios of 1.5°C, 3°C, and 5°C were selected, consistent with Mendelsohn and Neumann (1999).[3] Precipitation changes for these incremental scenarios were based on the average GCM outputs per degree of change in temperature as presented in Table 4.1 and Plates 1 and 2. In addition, we examined a set of scenarios where just temperature increases and precipitation does not change. The incremental scenarios are summarized in Table 4.2. Note that there was virtually no change in summer precipitation because the meteorological conditions that drive California's summer climate (that is, persistent high atmospheric pressure and cold ocean temperatures) are mostly unchanged in the model simulations.

The GCMs provide more spatial and temporal variance, but are quite data intensive. For example, the GCMs give different estimates of climate change for each season for each region in California. The uniform scenarios assume

Table 4.2   California 2100 incremental climate change scenarios

Incremental scenarios	
Temperature (°C)	Percent change in precipitation
1.5	0
1.5	9
3	0
3	18
5	0
5	30

the same change in each season and across the entire state and so are easy to develop and interpret, but lack the spatial and temporal differences that the GCMs estimate. Including both GCM and incremental uniform scenarios provides a rich set of alternative future climate outcomes upon which to base a regional analysis of impacts.

Ideally, we would incorporate into the analysis changes in climate variability, storms, or longer-term climate events such as El Niño or the PDO. However, although climate variability may change, information on which to base plausible changes in variability is currently insufficient. For example, the estimated changes in daily variance from the GCMs are not reliable and creating scenarios of changes in daily variance is computationally difficult (for example a change in daily rainfall amounts must equal a change in total monthly rainfall coming from a GCM or another source). One option would be to use a weather generator, but doing so for the entire state of California would be challenging because the scenario would have to be spatially correlated between the generated outputs across grid points. So, the assumption is made that only average conditions change. Note this can result in changes in intensity of extreme events, that is, an increase in average precipitation results in more precipitation events above a certain level, for example 1 inch per day. However, this approach does not capture the influence of changes in variance, storms, or long-term climate events.

The impact researchers relied on the two GCMs to provide decadal forecasts from 2000 through 2100. The average estimates of statewide changes in temperature and precipitation in the early, middle, and latter parts of the twenty-first century are displayed in Table 4.3. In some circumstances, the researchers also examined an intertemporal path generated from the incremental scenarios. For the incremental scenarios, it was assumed that

Table 4.3   GCM predictions for 2020, 2060, and 2100

Scenario	Temperature	Precipitation
Annual change 2010–39		
HadCM2	+1.4°C	+24.6%
PCM	+0.5°C	−6.5%
Annual change 2050–79		
HadCM2	2.4°C	+32.0%
PCM	1.6°C	−11.7%
Annual change 2080–99		
HadCM2	3.3°C	+58%
PCM	2.4°C	−21%

climate change was linear from 2000 through 2100. For example, with a 3°C change in 2100, the temperature change in 2050 would be 1.5°C. The scenarios were created by combining historical observations with the changes predicted by the scenarios. Two approaches were used. In the vegetation modeling, monthly changes from the GCMs were combined with current average monthly climate data for each downscaled grid box. This yields a transient of change in temperature and precipitation at a 100 km^2 scale. In the second approach, average changes in climate were calculated for 2010–39, 2040–69, and 2080–99. These averages were combined with observed data from 1963 to 1992 in the runoff study and observed data from 1971 to 2000 in the crop yield study. Note that in some applications, month by month results were used from the GCMs. This captures changes in intermonthly, interseasonal, and interannual variability as simulated by the GCMs.

While they do include a wide range of potential changes in climate, the scenarios examine only changes in average conditions. By assuming that only average conditions change, the intensity and frequency of extremes are also assumed to change. For example, by assuming a rise in temperature, then the hottest days will be even hotter and more days will be above a particular temperature threshold, for example 95°F. It would also result in fewer frost days. An assumed increase in average precipitation would result in those days with precipitation having a greater amount. This results in the number of days with precipitation above a certain amount increasing. Assuming a reduction in precipitation would have the opposite effect.

Assumed carbon dioxide concentrations (Wigley, 2000) are shown in Table 4.4. These concentrations reflect an assumed 'business as usual' emissions scenario, similar to the Intergovernmental Panel on Climate Change's IS92a scenario, in which rates of growth of emissions as well as improvements in energy efficiency, social development, etc., are assumed to follow current trends (Leggett et al., 1992).

Even though these scenarios cover a broad range of potential changes in climate, more work is needed to develop scenarios with high spatial resolution, and scenarios that capture potential changes in multi-year climate events (such as El Niño events) and in daily to monthly events (such as

Table 4.4 Assumed CO_2 concentrations by decade

Year	CO_2 concentrations (ppm)
2020	417
2060	543
2100	712

storms and droughts). Also, decision makers would find the scenarios much more useful if we could reasonably assign probabilities to them instead of treating them as equally plausible.

NOTES

1. In a related study (http://www.energy.ca.gov/reports/2003-10-31_500-03-058CF_A14. PDF), the Electric Power Research Institute (EPRI) examined changes in hospital admissions in California during El Niño episodes.
2. The PCM scenario output is not included in Table 4.1.
3. The middle change in temperature of 3°C is consistent with a global average warming of 2.5°C, which is considered to be the most likely climate change percentage by 2100 (Wigley, 1999). On average, the GCMs showed that California will warm 14 percent more than the global average climate. Applying the 14 percent to a 2.5°C increase in global average temperature yielded an approximate 3°C increase in California temperature.

REFERENCES

Dai, A., G.A. Meehl, W.M. Washington, T.M.L. Wigley, and J.A. Arblaster. 2001. Ensemble simulation of 21st century climate changes: Business as usual versus CO_2 stabilization. *Bulletin of the American Meteorology Society* **82**: 2377–88.

Daly, C., T.G.F. Kittel, A. McNab, J.A. Royle, W.P. Gibson, T. Parzybok, N. Rosenbloom, G.H. Taylor, and H. Fisher. 1999. Development of a 102-year high resolution climate data set for the conterminous United States. In *Proceedings, 10th Symposium on Global Change Studies*, 79th Annual Meeting of the American Meteorological Society, 10–15 January, Dallas, TX, pp. 480–83.

Fu, C.B., S.Y. Wang, Z. Xiong, W.J. Gutowski, D.K. Lee, J. McGregor, Y. Sato, H. Kato, J.-W. Kim, M.-S. Su. 2005. Regional climate model intercomparison project for Asia. *Bulletin of the American Meteorological Society*. **86**(2): 257–66.

Johns, T.C., R.E. Carnell, J.F. Crossley, J.M. Gregory, J.F.B. Mitchell, C.A. Senior, S.F.B. Tett, and R.A. Wood. 1997. The second Hadley Centre Coupled Ocean-Atmosphere GCM Model description, spinup, and validation. *Climate Dynamics* **13**: 103–4.

Kerr, R.A. 2001. A little sharper view of global warming. *Science* **294**: 765.

Leggett, J., W.J. Pepper, and R.J. Swart. 1992. Emissions scenarios for IPCC: An update. In *Climate Change 1992 – The Supplementary Report to the IPCC Scientific Assessment*. J.T. Houghton, J.T., B.A. Callander, and S.K. Varney (eds). WMO/UNEP Intergovernmental Panel on Climate Change. Cambridge University Press, Cambridge, UK, pp. 69–95.

Mendelsohn, R. and J. Neumann (eds). 1999. *The Impact of Climate Change on the US Economy*. Cambridge University Press, Cambridge, UK.

NAST. 2000. *Climate Change Impacts on the United States: The Potential Consequences of Climate Variability and Change*. National Assessment Synthesis Team, US Global Change Research Program, Washington, DC.

Trenberth, K.E. and T.J. Hoar. 1996. The 1990–1995 El Niño/Southern Oscillation event: Longest on record. *Geophysical Research Letters* **23**(1): 57–60.

Wigley, T.M.L. 1999. *The Science of Climate Change: Global and US Perspectives.* The Pew Center on Global Climate Change, Arlington, VA. Available at www.pewclimate.org.

Wigley, T.M.L. 2000. Stabilization of CO_2 concentration levels. In *The Global Carbon Cycle*, T.M.L. Wigley and D.S. Schimel (eds). Cambridge University Press, Cambridge, UK, pp. 258–76.

Western Regional Climate Center. 2005. Climate of California. http://www.wrcc. dri.edu/narratives/CALIFORNIA.htm. Accessed 8 February 2005.

5. Terrestrial ecosystem changes

James M. Lenihan, Raymond Drapek, and Ronald Neilson

5.1 INTRODUCTION

California is one of the most climatically and biologically diverse areas in the world. There is more diversity in the state's land forms, climate, ecosystems, and species than in any comparably sized region in the United States (Holland and Keil, 1995). The diversity of landscapes and climates supports a broad range of natural ecosystems ranging from the cool and wet redwood forests of the northwestern bioregion to the hot and dry Mojave and Sonoran deserts (Hickman, 1993). This diversity of habitats sustains a greater level of species diversity and endemism than is found in any other region of the nation. The California flora constitutes approximately 25 percent of the flora of the continental United States, and about 25 percent of the plant species are endemic to the state (Davis et al., 1998).

California's natural systems support many important economic activities. Its forests provide 10 percent of the nation's wood products, livestock grazing ranks second among all agricultural commodities in the state, and more than 55 percent of the state is used for logging, grazing, or both (Jensen et al., 1993). The 11 million acres of parks and wilderness areas draw millions of tourists to the state each year, more than any other state except Alaska (California Division of Tourism, 2001). In addition, the natural systems provide numerous ecosystem services (for example water and air purification, mitigation of droughts and floods, cycling and movement of nutrients, and control of potential agricultural pests) that help sustain many sectors of California's economy (Field et al., 1999).

The state's burgeoning population and the consequent impacts on the landscape threaten much of its biological wealth. Throughout the state, natural habitats have been and continue to be altered and fragmented, endangering the state's biological diversity (Barbour et al., 1993). Most of California's forests have been logged, native oak woodlands are in serious decline, native grasslands have almost completely disappeared, and nearly 90 percent of the wetlands and riparian areas have been severely degraded or destroyed. Even relatively unmanaged natural systems have been significantly altered by the introduction of non-native species and by fire suppression (Field et al., 1999). Pervasive human impacts, along with the high level of species endemism, have placed a large number of species in California at risk of extinction. The 227 species currently listed as endangered or threatened in the state is a higher number than in any other state (US Fish and Wildlife Service, 2001).

In the future, global climate change will increasingly interact with and intensify the pressures of a growing population on California's natural ecosystems. Recent studies show that even gradual and apparently small changes in climate can lead to catastrophic shifts in ecosystems when ecosystem resilience has already been compromised by human exploitation (Scheffer et al., 2001). Regional climate studies indicate that, on average, California may experience substantially warmer and wetter winters, somewhat warmer summers, and an enhanced El Niño/Southern Oscillation (ENSO) during the next century (Field et al., 1999; Gutowski et al., 2000). All natural ecosystems, whether managed or unmanaged, will be affected by these changes in climate. It is not possible to predict accurately the response of the natural systems to global climate change through direct experimentation. The physical extent, complexity, and expense of even a single-factor experiment for an entire ecosystem make such an experiment usually prohibitive (Aber et al., 2001). However, analyses of the sensitivity of natural ecosystems to climate change can be made using ecosystem models that integrate information from direct experimentation.

In this study, we used MC1 (Daly et al., 2000; Bachelet et al., 2001b; Lenihan et al., 2003), a state-of-the-art dynamic vegetation model, to investigate the sensitivity of natural ecosystems in California under two different future climate scenarios.[1] MC1 simulates vegetation succession at large spatial extents through time while estimating variability in the carbon budget and responses to episodic events such as drought and fire. Although MC1 does not as yet simulate interactions with land use effects or constraints on ecosystem change imposed by dispersal of propagules, the model has been used to examine the sensitivity of natural ecosystems to global climate change for several national-scale studies, most recently for the US Global Change Research Program's National Assessment of Climate Change Impacts on the United States (Aber et al., 2001; National Assessment Synthesis Team, 2001). Because MC1 is capable of predicting the dynamic path of ecosystem change, it represents a significant improvement over past models that could represent only changes in equilibrium states. The applications of the ecosystem model to endangered species in Chapter 6 and timber in Chapter 7 both take advantage of these dynamic forecasts.

5.2   METHODS

5.2.1   The Model

MC1 (Daly et al., 2000; Bachelet et al., 2001b) is a dynamic vegetation model that simulates life-form mixtures and vegetation types; ecosystem fluxes of carbon, nitrogen, and water; and fire disturbance (Lenihan et al., 1998). MC1 is routinely implemented (Bachelet et al., 2000, 2001a; Daly et al., 2000; Aber et al., 2001) on spatial data grids of varying resolution (that is, grid cell sizes ranging from 30 m^2 to about 2500 km^2) where the model is run separately for each grid cell (that is, there is no exchange of information across cells). The model reads climate data at a monthly timestep and runs interacting modules that simulate biogeography, biogeo-chemistry, and fire disturbance.

5.2.1.1   Biogeography module
The biogeography module simulates the potential life-form mixture of evergreen needleleaf, evergreen broadleaf, and deciduous broadleaf trees, along with C$_3$ and C$_4$ grasses. The tree life-form mixture is determined at each annual timestep by locating the grid cell on a two-dimensional gradi-ent of annual minimum temperature and growing season precipitation. Lifeform dominance is arrayed along the minimum temperature gradient from more evergreen needleleaf dominance at relatively low temperatures, to more deciduous broadleaf dominance at intermediate temperatures, to more broadleaf evergreen dominance at relatively high temperatures. The precipitation dimension is used to modulate the relative dominance of deciduous broadleaved trees, which is gradually reduced to zero toward low values of growing season precipitation. Mixtures of C$_3$ versus C$_4$ grasses are determined by reference to their relative potential productivity during the three warmest consecutive months. Potential grass production by life-form is simulated as a function of soil temperature using equations from the CENTURY model (Parton et al., 1994). The tree and grass life-form mixtures, together with leaf biomass simulated by the biogeochemistry module, are used in a rule base to determine which of 22 possible potential vegetation types occurs at the grid cell each year. The MC1 biogeography rule base was developed using the MAPSS (Neilson, 1995) rule base as a template.

5.2.1.2   Biogeochemistry module
The biogeochemistry module is a modified version of the CENTURY model (Parton et al., 1994), which simulates plant productivity, organic matter decomposition, and water and nutrient cycling. Plant productivity is

constrained by temperature, effective moisture (that is, a function of soil moisture and potential evapotranspiration), and nutrient availability. The simulated effect of increasing atmospheric carbon dioxide (CO_2) is to increase maximum potential production and to decrease transpiration (thus reducing the constraint of effective moisture on productivity). Trees compete with grasses for soil moisture, light, and nutrients. Competition for water is structured by rooting depth. Trees and grasses compete for soil moisture in the upper soil layers, where both lifeforms are rooted, but the trees with deeper roots have sole access to moisture in deeper soil layers. Grass productivity is constrained by light availability in the understory; the light availability is reduced as a function of tree leaf carbon. Parameterization of the tree and grass growth processes in the model is based on the current lifeform mixture, which is updated annually by the biogeography module. For example, an increase in annual minimum temperature that shifted the dominance of evergreen needleleaf trees to co-dominance with evergreen broadleaf trees would trigger an adjustment of tree growth parameters (for example the optimum growth temperature) that would, in turn, produce a modified tree growth rate.

5.2.1.3   Fire disturbance module

The MC1 fire module (Lenihan et al., 1998) simulates the occurrence, behavior, and effects of fire. The module consists of several mechanistic fire behavior and effect functions (Rothermel, 1972; Peterson and Ryan, 1986; van Wagner, 1993; Keane et al., 1997) embedded in a structure that provides two-way interactions with the biogeography and biogeochemistry modules. Live crown structure and fuel loading in several size classes of both dead and live fuels are estimated using life-form-specific allometric functions of the different carbon pools. The moisture content of each dead fuel size class is estimated as a function of antecedent weather conditions averaged over a period of days dependent on size class. The moisture content of each live fuel size class is a function of the soil moisture content to a specific depth in the profile. Fuel moisture and distribution of the total fuel load among different size classes determine potential fire behavior estimated using the Rothermel (1972) fire spread equations.

The rate of fire spread and the intensity of the fire line are the model estimates of fire behavior used to simulate fire occurrence and effects. A fire event is triggered by thresholds of fire spread, fine fuel flammability, and coarse woody fuel moisture (given a constraint of just one fire event per year). The thresholds were calibrated to limit the occurrence of simulated fires to only the most extreme events. Large and severe fires account for a very large fraction of the annual area burned historically (Strauss et al., 1989). These events are also likely to be least constrained by heterogeneities in topography, fuel moisture, and fuel loading that are poorly represented by relatively coarse-scale input data grids (Turner and Romme, 1994).

The direct effect of fire in the model is the consumption and mortality of dead and live vegetation carbon that is removed from (or transferred to) the appropriate carbon pools in the biogeochemistry module. This direct effect is a function of the simulated fraction of the cell burned, fire-line intensity, and tree canopy structure. The fraction of the cell burned depends on the simulated rate of fire spread and the time since the last fire event relative to the current fire return interval simulated for the cell. Higher rates of spread and longer intervals between fires generally produce more extensive fire events in the model. Live carbon mortality and consumption within the burned area are functions of fire-line intensity and the tree canopy structure (that is, crown height, crown length, and bark thickness). Dead biomass consumption is simulated using functions of fire intensity and fuel moisture that are fuel-class specific.

Fire effects extend beyond the direct impact on carbon and nutrient pools to more indirect and complex effects on tree versus grass competition. Fire tends to tip the competitive balance toward grasses in the model because much, or all, of the grass biomass consumed regrows in the year after a fire. Woody biomass consumed or killed is more gradually replaced. A greater competitive advantage over trees promotes greater grass biomass. This, in turn, produces higher fine fuel loadings and changes in the fuel bed structure that promote greater rates of spread and thus more extensive fire. In contrast to the simulated rate of spread, which is largely dependent on fine fuel properties, fire-line intensity is more dependent on the properties of the total fuel load. Thus, an increase in tree biomass, which contributes more to the heavy coarse fuels, promotes more intense fire and more biomass consumption and mortality. This, in turn, acts as another negative effect on tree biomass. However, increases in tree biomass also reduce the productivity of grasses, which reduces both fine fuel and total fuel loadings, adding another layer of complexity to the fire–vegetation interactions in the model.

5.2.2   Climatic Data

The climate data used as input to the model in this study consisted of monthly time series for all the necessary variables (that is, precipitation, minimum and maximum temperature, and vapor pressure) distributed on a 100 km^2 resolution data grid for California as described in Chapter 4. This study also applied the incremental scenarios.

5.3 RESULTS AND DISCUSSION

5.3.1 Simulation Results for the Historical Climate

5.3.1.1 Vegetation classes

Of the 22 possible vegetation types predicted by the biogeography module, 12 occurred in the simulations for California. These types were aggregated into seven vegetation classes to simplify the visualization of results. The aggregation scheme and lists of typical regional examples in each vegetation class (Table 5.1) indicate the range of each class in terms of physiognomy and species dominance.

The simulated vegetation class distribution is difficult to validate against actual vegetation for California. The MC1 biogeography module simulates the life-form mixture and vegetation type that could potentially occur given climatic conditions and the simulated fire regime. However, the actual distribution of vegetation types has been highly modified by urbanization, agriculture, and forestry practices, including fire suppression. Nonetheless, the overall distribution of the vegetation classes simulated for the historical period (Plate 3) was very similar to that shown on published vegetation type maps of California.[2]

5.3.1.2 Carbon density

Average carbon density values for the simulated vegetation classes compared favorably to observed mean carbon density values derived for equivalent vegetation classes (Atjay et al., 1979; Houghton and Hackler, 2000). The distribution of simulated average total ecosystem carbon (Plate 4a) and average total vegetation carbon (Plate 4b) for the 30 year historical period also appears to be relatively accurate, with the highest carbon density in the most heavily forested regions of the state (the Northwest, Cascade Range, and Sierra Nevada regions), the lowest carbon density in the most arid regions (the Mojave and Sonoran deserts), and intermediate values throughout the rest of the state. This carbon density gradient is even more distinct in the distribution of live tree carbon where the highest values define the forested portion of the state. The distribution of grass carbon density is negatively related to the distribution of woody carbon density, and the highest grass values are simulated in regions of the state (for example the Central Valley and surrounding foothills, southern coast, and intermountain basin) where grassland likely was a dominant element of pre-settlement vegetation (Holland and Keil, 1995).

5.3.1.3 Fire regime

The MC1 fire module simulates fire severity under conditions of potential natural vegetation and no fire suppression, so validating the historical

Table 5.1   MC1 vegetation type aggregation scheme and regional examples of the vegetation classes

MC1 vegetation class	MC1 vegetation type	Regional examples
Alpine/subalpine forest	Tundra	Alpine meadows
	Boreal forest	Lodgepole pine forest
		Whitebark pine forest
Evergreen conifer forest	Maritime temperate conifer forest	Coastal redwood forest
		Coastal closed-cone pine forest
	Continental temperate coniferous forest	Mixed conifer forest
		Ponderosa pine forest
Mixed evergreen forest	Warm temperate/ subtropical mixed forest	Douglas fir/tan oak forest
		Tan oak/madrone/ oak forest
		Ponderosa pine/black oak forest
Mixed evergreen woodland	Temperate mixed xeromorphic woodland	Blue oak woodland
		Canyon live oak woodland
	Temperate conifer xeromorphic woodland	Northern juniper woodland
Grassland	C_3 grassland	Valley grassland
	C_4 grassland	Southern coastal grassland
		Desert grassland
Shrubland	Mediterranean shrubland	Chamise chaparral
	Temperate arid shrubland	Southern coastal scrub
		Sagebrush steppe
Desert	Subtropical arid shrubland	Creosote brush scrub
		Saltbrush scrub
		Joshua tree woodland

simulation of fire involves some of the same difficulties described for the vegetation class simulation. The simulated distribution of the fire rotation period (that is, the expected time to burn an area the size of the grid cell; Heinselman, 1973), which was calculated for the 100 years of historical climate (Plate 5a), showed that, for more than 90 percent of the state, the rotation period was about 6–20 years. Taking the fire rotation period as an estimate of point fire frequency (Lertzman et al., 1998), this result compares favorably with observed point-based frequencies of 5–10 years in woodlands and grasslands and 4 to 20 years in mixed and conifer forests of California (Skinner and Chang, 1996). Considerably longer mean fire rotation periods of 20–100 years were simulated for coastal forests and shrublands where fire activity is constrained by relatively high humidity and fuel moisture, and in the most arid portions of the state where fire activity is constrained by relatively low fuel loads.

The thresholds of drought and fine fuel flammability that trigger fire events in the MC1 fire module are set to constrain events to those of relatively high intensity and severity. The distribution of fire-line intensity (Plate 5b) averaged over all fire events in the 100-year historical simulation is consistent with the distribution of vegetation classes and their characteristic fire behaviors under severe burning conditions. Rothermel (1983) provides a classification of fire-line intensities characteristic of surface fire ($<$ 500 Btu/ft/s), mixed surface and passive crown fire (500–1000 Btu/ft/s), and active crown fire ($>$ 1000 Btu/ft/s). By this measure, the historical intensities simulated by MC1 are primarily in the range of crown fire in the forest classes, the mixed fire regime in the woodlands and shrubland classes, and surface fire in the grasslands and desert classes.

Martin and Sapsis (1995) estimated that fire burned 5.5 percent to 13.0 percent of California annually under pre-settlement conditions. The range of percentage of area burned simulated by MC1 during the 100 years of the historical simulation (6.3 percent–15.5 percent) was remarkably similar to this independent estimate. The simulated trend of percentage of area burned showed a significant and fairly strong relationship (Spearman rank correlation $= -0.70$, $p < 0.001$) with the historical trend of the Palmer's Z drought index for California. Statewide values of the index for the period of record were calculated by averaging over values for all five climatic divisions in California (Karl, 1986). Several of the most severe fire years simulated by the model were coincident with some of the most severe drought years (for example 1910, 1924, 1928, 1959, and 1966), and several of the least severe fire years correspond to some of the wettest years (for example 1906, 1912, 1941, 1958, and 1983). Another pattern evident in the relationship between simulated fire and the drought index was the occurrence of a wet episode for one or two years before a dry and severe fire year (for example 1906 and 1908, 1958 and 1959, 1982 and 1983, and 1984). Several of the most severe fire years (for example 1908, 1959, and 1984) were preceded by one or more relatively wet years in which a build-up of fuels was simulated by the model. A similar pattern of large fire years promoted by sequences of wet seasons followed by average or drier than average seasons has been identified in the southwestern United States (Swetnam and Betancourt, 1998).

In an attempt to validate the simulated trend of area burned during the historical period, the simulation results were compared to the observed trend on US national forest lands within the Sierra Nevada bioregion during the 1908–93 period (McKelvey and Busse, 1996). The simulated mean annual area burned for this region during the historical period (87 kha/yr) was significantly greater than the observed mean (17 kha/yr). It is likely that fire suppression and ignition constraints have reduced the observed mean annual area burned to some unknown level below the potential value estimated by the model (Husari and McKelvey, 1996; McKelvey and Busse, 1996). The observed and simulated trends showed a significant but only moderately strong correlation (Spearman rank correlation coefficient = 0.52, $p < 0.001$). The observed data does validate the occurrence of several severe fire years simulated by the model (for example 1910, 1917, 1924, 1928, 1939, 1959, and 1987).

5.3.2   Simulations for the Future Climate Scenarios

5.3.2.1   Vegetation classes
The response of vegetation class distribution under the two future climate scenarios was determined by comparing the distribution of the most frequent vegetation type simulated for the 30-year historical period (Plate 3a) against the same for the last 30 years (2071–99) of the future scenarios (Plate 3b, 3c). The simulated response of the vegetation classes in terms of changes in percentage of coverage (Figure 5.1) was surprisingly similar under the two future climates. There was agreement on the direction of change (that is, decrease or increase in coverage) for all but the evergreen conifer forest class, and the amounts of change were comparable for a few of the vegetation classes. However, these similarities in the response of class coverage were often the net result of very different responses to each scenario in terms of the spatial distribution of vegetation classes.

Hadley scenario   A prominent feature of the response of the vegetation class distribution under the Hadley scenario (Plate 3b) was the advancement of forest classes into the Modoc Plateau, into the northern end of the Great Central Valley, toward higher elevations in the Sierra Nevada, and inland along the coast. Increases in both temperature and moisture under this scenario favored expansion of forest, and they were especially favorable for mixed evergreen forest. The relatively high degree of warming under the Hadley scenario promoted a widespread change in the simulated life-form composition from needleleaf dominance to mixed needleleaf–broadleaf in the northern half of the state. Consequently, mixed evergreen forest replaced evergreen conifer forest throughout much of the latter's simulated historical range. Two examples of this transition in terms of species dominance within the different bioregions might include the replacement of

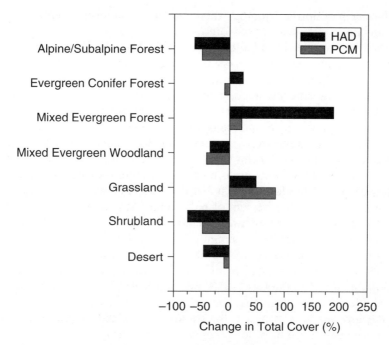

Figure 5.1 Percent changes in the total cover of the vegetation classes under the Hadley and PCM scenarios

Douglas fir–white fir forest by Douglas fir–tan oak forest in the northwest bioregion, and the replacement of white fir–ponderosa pine forest by ponderosa pine–black oak forest in the Sierra Nevada. Greater moisture availability under the Hadley scenario also promoted the advancement of mixed evergreen forest into mixed evergreen woodland, shrubland, and grassland. Movement into the northern end of the Great Central Valley could represent the replacement of blue oak woodlands, chaparral, and perennial grassland by tan oak–madrone–canyon live oak forest with scattered Douglas fir and ponderosa pine. In the central western and southwestern bioregions, mixed evergreen forests of coast live oak–madrone or canyon live oak–Coulter pine might replace chaparral and live oak woodlands.

Evergreen conifer forest showed a net increase in percentage of coverage under the Hadley scenario despite the loss of much of its simulated historical range to mixed evergreen forest. The main region of evergreen conifer advancement was in the cold desert region of the Modoc Plateau and east of the Sierra Nevada. Here higher moisture availability promoted the advancement of evergreen conifer forest into mixed evergreen woodland and shrubland. On the Modoc Plateau, this transition would be characterized by replacement of northern juniper woodland and Great Basin sagebrush scrub by ponderosa pine–Jeffrey pine forest.

Maritime evergreen conifer forest, one of the MC1 vegetation types in the evergreen conifer vegetation class, is distinguished in the rule base by very high tree leaf carbon density and a low index of continentality (that is, difference between highest mean monthly temperature and lowest mean monthly temperature) simulated for the relatively moist and equable climate along the northern and central coast of California. In the historical simulation, these criteria effectively distinguished the conifer-dominated forests (for example redwood and closed-cone pine forests) along the north-central coast, where effective moisture is high and temperatures are most equable. Advancement of this forest type under the wetter and warmer Hadley scenario would represent movement of redwood forest inland into the Douglas fir–tan oak forest in the northwestern bioregion, and expansion of redwood and closed-cone pines from remnant, fragmented groves into surrounding canyon live oak–madrone forests in the central western region.

Evergreen conifer forest also advanced into the high-elevation subalpine and alpine forest in the Cascade Range and Sierra Nevada regions under the Hadley scenario. Here the model responded to an increase in the length of the growing season past a threshold in the biogeographic rule base. Advancement of red fir or lodgepole pine forest into subalpine parks and meadow would be an example of this transition.

In addition to widespread advancement of forest, another prominent feature of the response of vegetation distribution under the Hadley scenario was the advancement of grassland, particularly in the southern end of the Great Central Valley and in the uplands of the Mojave Desert where grassland replaces desert. Here the response to increased precipitation was an increase in both tree and grass biomass. More fine fuels supported more fire (Plate 5a) which favored grasses. In the Mojave Desert, this transition could represent an increase in the extent of the desert grasslands interspersed with Joshua tree desert woodland and creosote bush scrub.

Parallel climate model scenario   The most prominent feature of the vegetation class response to the drier PCM scenario (Plate 3c) was the advancement of grassland into the simulated historical range of mixed evergreen woodland and shrubland. This transition was prompted by a decline in the competitiveness of woody life-forms relative to grasses as a response to a decline in effective moisture, and an increase in fire, which further constrained woody lifeform production. The advancement of grassland occurred primarily on the Modoc Plateau, in the foothills surrounding the Great Central Valley, and in the interior of the central western region. On the Modoc Plateau, a likely example of this transition would be an increase in the extent of the grassland that is interspersed within the northern juniper woodland and sagebrush scrub communities. A similar transition is already occurring under present-day conditions in the sagebrush scrub communities of the intermountain West. Here drought, increasing cheatgrass abundance, and fire are interacting to significantly reduce the woody scrub component (D'Antonio and Vitousek, 1992). In the foothills of the Great Central Valley and the central western bioregions, the model simulation could indicate the loss of various oak woodland and chaparral communities to non-native grassland advancement.

Mixed evergreen woodland and shrubland show too little advancement to compensate for the retreat from grassland under the PCM scenario. Consequently, a net decline is seen in the coverage of mixed evergreen woodland and shrubland (Figure 5.1), along with a narrowing of the simulated ecotones between forest and grassland (Plate 3a, 3c). One local exception to this trend is in the eastern half of the northwest region, where there was some mixed evergreen woodland advancement into forest. In this bioregion, northern oak woodland advancing into Douglas fir–tan oak and Douglas fir–white fir forests would be a likely example of this transition. An exception to the general decline in shrubland was in the Sierra Nevada region, where shrubland advanced into alpine and subalpine forest. Here a regional example of the model's response to a lengthened growing season could be an increase in whitebark pine krummholtz within alpine meadow communities.

In contrast to the simulation for the Hadley scenario, the distribution of mixed evergreen forest and evergreen conifer forest remained relatively static under the PCM scenario. Mixed evergreen forest showed a relatively small gain in coverage (Figure 5.1) with limited advancement into evergreen conifer forest in the northwestern and Sierra Nevada bioregions. This transition was prompted by a temperature-driven shift in life-form composition similar to that seen under the Hadley scenario, but the response was more constrained under the cooler PCM scenario. There was a net loss in the statewide coverage of evergreen conifer forest (Figure 5.1), but the class showed some advancement at high elevations in the Sierra Nevada and on the Modoc Plateau, and along the coast in the central western region. In these relatively cool regions of the state, tree productivity increased as a response to increases in temperature and relatively small declines in precipitation.

Incremental scenarios   The common response of the vegetation classes under all four incremental scenarios was an increase in the total coverage of mixed evergreen forest and grasslands and a decrease in the coverage of the other vegetative types. The mixed evergreen forest responded to the warmer temperatures by advancing into the historical range of evergreen conifer forests. This response to temperature was the result of a widespread change in the simulated life-form composition from needleleaf dominance to mixed needleleaf–broadleaf in the northern half of the state. Mixed evergreen forests also responded to the wetter scenarios (T3P18 and T5P30) by advancing into the historical range of mixed evergreen woodlands.

Grassland responded to the drier scenarios (T3P0 and T5P0) by advancing into areas currently occupied by mixed evergreen woodlands and shrublands. The decrease in effective moisture favored grass over woody lifeforms at the ecotones between grasslands and these two vegetation classes. Grasslands responded to the wetter scenarios (T3P18 and T5P30) by advancing into deserts. Increased available moisture at the ecotone between these two classes prompted increases in both grass and woody carbon density. The increase in total vegetation biomass produced more fuel and fire, which favored the expansion of grasslands.

Losses in the coverage of evergreen conifer forests to the advancement of mixed evergreen forests under all incremental scenarios were partially counterbalanced by gains in the Modoc Plateau and eastern Sierra Nevada bioregions. Here warmer and especially wetter conditions prompted an increase in woody carbon density and a resultant expansion of evergreen conifer forests into woodlands and shrublands. The net result of this retreat and advancement of evergreen conifer forests ranged from nearly no change in coverage under the most cool and wet scenario (T3P18) to the greatest decrease under the most warm and dry scenario (T5P0).

Mixed evergreen woodlands and shrublands were uniformly reduced under all the incremental scenarios. Losses under the drier scenarios (T3P0 and T5P0) were mostly to the advancement of grasslands, and losses under the wetter scenarios (T3P18 and T5P30) were typically to the advancement of the mixed evergreen and evergreen conifer forests.

The losses in coverage of deserts under all the incremental scenarios were to expansions in grasslands. Under the drier scenarios (T3P0 and T5P0) these losses were relatively small, and confined to the southern end of the Great Central Valley, where the frequency of grasslands and deserts are estimated to be nearly equal in the last 30 years of the twenty-first century. Under the wetter scenarios (T3P18 and T5P30), deserts suffer relatively greater losses to grasslands in the southern Central Valley and within the Mojave and Sonoran desert bioregions.

The losses in coverage of the alpine and subalpine forests to advancement of evergreen conifer forests and shrublands were predominately a response to a longer growing season brought on by the warmer temperatures, and were most pronounced under the warmest scenarios (T5P0 and T5P30).

5.3.2.2 Carbon density and fire regime

Hadley scenario Underlying the advance of the forest classes under the Hadley scenario was an even more widespread increase in tree carbon density. Tree biomass increased as a response to increased total annual precipitation and the amplification of the winter wet and summer dry seasonal cycle, both of which favored increased tree growth. The amount and distribution of tree biomass were strongly related to total annual precipitation. For example, the 500 mm isoline for the total annual precipitation was a consistently accurate predictor of tree leaf biomass at the threshold level between the woody vegetation classes and grassland in the simulations for the historical and future climate periods. And in the Hadley scenario, precipitation falling in months outside the growing season was an even greater proportion of the total than had been seen historically, and this favored trees over grasses in the simulated competition for water. Precipitation during non-growing season months is particularly effective at recharging the moisture of the deeper soil layers, thus differentially promoting the growth of the deeper rooted trees over grasses. Increased tree leaf biomass reduced light availability in the understory, which constrained grass growth and increased the moisture and nutrients available to trees.

The biomass of both trees and grass increased under the Hadley scenario as a response to increased precipitation in the southern half of the Great Central Valley and in the highlands of the Mojave Desert. The increase in total vegetation biomass (Plate 4b), and especially the increase in grass biomass, produced more fire (that is, shorter mean fire rotation periods, Plate 5a) and higher fire intensity (Plate 5b). A relatively small area of tree carbon decline was simulated under the Hadley scenario, and the decline was located primarily in the most arid portion of the state in the Sonoran Desert and the lowlands of the Mojave Desert. Here greater grass biomass supported by increased precipitation also produced more frequent and extreme fire (Plate 5a, 5b), which helped constrain woody biomass.

The simulated trend in percentage of area burned under the Hadley scenario (Figure 5.2a) was more variable than the simulated historical trend (coefficients of variation were 21 percent and 15 percent). The increase in precipitation under the Hadley scenario increased the variability of the fire regime by reducing the area burned to lower levels and contributing to greater biomass build-up during the relatively wet years, thus setting the system up for higher levels of burned area promoted by higher fuel loads during the relatively dry years. This interaction between fuels and interannual variability in precipitation produced the somewhat counter-intuitive result of more severe fire years simulated under the wetter Hadley scenario. The model predicts three fire events (in 2027, 2044, and 2074) under the

Hadley scenario (Figure 5.2a) that are more severe than any event simulated for the historical period. The smoothed trend in simulated percentage of area burned for the Hadley scenario (Figure 5.2b) shows a distinct rise above the simulated historical mean in the last few decades of the future period. However, linear regression analysis showed that, overall, the slope of the Hadley trend was not significantly different than zero.

PCM scenario   In contrast to the response under the Hadley scenario, there was a widespread decrease in tree carbon density in response to the decrease in total annual precipitation under the PCM scenario. The reduction in deep soil moisture recharge shifted the competitive advantage to grasses, producing a widespread increase in grass biomass that was strongly related to the decline in tree biomass. Fire frequency increased (Plate 5a) where the increase in grass biomass was greatest, and fire intensity increased (Plate 5b) where the net result of increased grass and decreased tree biomass was a gain in total vegetation biomass (Plate 4b).

Two exceptions to this widespread response to the PCM scenario were along the coast and along the Sierra Nevada, where the biomass of trees increased and grasses declined. In these relatively cool parts of the state, simulated tree growth was restricted by temperature and moisture availability. Decreases in precipitation were relatively small under the PCM scenario in these two regions. Here increased temperatures without significant changes in moisture availability promoted increased tree growth and corresponding declines in grass growth. The decline in grass biomass produced a reduction in fire frequency along the coast, where it was already very low in the historical simulation (Plate 5a). In fact, fire frequency dropped to zero or near zero along much of the coast under this scenario, producing the large percentage changes in fire intensity (Plate 5b).

The most arid portion of the state (the Sonoran Desert and the Mojave Desert lowlands) was the other region where an exception was seen to the general trend of increased grass biomass and decreased tree biomass under the PCM scenario. Here tree and grass biomass both declined. In this same region, the largest increases in annual temperature and the largest decreases in annual precipitation were predicted under the PCM scenario. The response to the decrease in effective moisture was a decrease in total vegetation biomass and attendant decreases in measures of simulated fire activity (Plate 5).

The simulated trend in percentage of area burned under the PCM scenario (Figure 5.2a) was less variable than the simulated historical trend (coefficients of variation were 10 percent and 15 percent). The smoothed trend in simulated percentage of area burned for the PCM climate scenario (Figure 5.2b) shows a distinct rise above the simulated historical mean in

A.

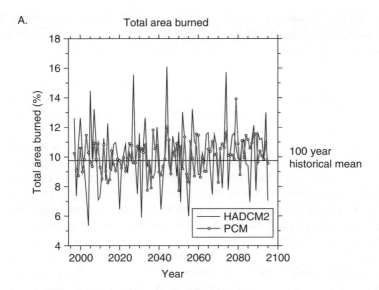

B.

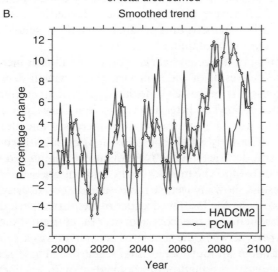

*Figure 5.2   Simulated trends in (a) the annual percentage of the total area
              burned and (b) the smoothed percentage deviation from the
              100-year historical mean for the future period (1994–2099) of
              the Hadley and PCM climate scenarios*

the last few decades of the future period, and linear regression analysis showed that the slope of the PCM trend was significantly different than zero.

Incremental scenarios   There was an increase over historical levels in average total ecosystem carbon density during the last 30 years under all four incremental scenarios. The increase ranged from 2.7 percent under the relatively warm and dry scenario (T5P0) to 5.2 percent under the relatively cool and wet scenario (T3P18). These increases were to a large extent due to changes in soil and litter carbon, which showed the greatest increase under the relatively cool scenarios (3C) and significantly less increase under the relatively warm scenarios (5C). The response of the soil and litter pool to changes in precipitation was relatively slight, suggesting the main response of this pool under the different incremental scenarios involved temperature-mediated changes in the rates of decomposition.

There was an increase over historical levels in average total live vegetation density during the last 30 years under all four incremental scenarios. This increase was especially pronounced under the two wetter scenarios (T3P18 and T5P30), where there were increases in both wood and grass carbon density. Declines in live wood carbon density under the drier scenarios (T3P0 and T5P0) were partially counterbalanced by even greater increases in grass carbon density, thus producing relatively small net gains in the total live vegetation pool under these two scenarios.

In summary, gains in total ecosystem carbon under all four incremental scenarios were the net result of gains in vegetation carbon density that more than compensated for losses associated with temperature-driven increases in the rates of soil and litter decomposition. Gains in live vegetation carbon density were a response to both increases and declines in effective moisture. Relatively small net gains in vegetation carbon were maintained even with declines in effective moisture via simulated shifts in life-form dominance toward the more drought-tolerant grasses.

Although the results show an average increase in carbon density relative to historical levels during the last few decades of the future period, trends in net biological production of the ecosystem over the entire 100 years show there were earlier periods when the simulated ecosystem was a source, not a sink, for carbon under all four incremental scenarios.[3] These periods of decline in net biological production are coincident with declines in the trend of total annual precipitation. Coupled with continued increases in mean annual temperature, these declines in effective moisture reduced vegetation productivity (and especially woody vegetation productivity). Subsequent reductions in live vegetation turnover to the soil and litter carbon pool, together with increasing rates of decomposition with higher

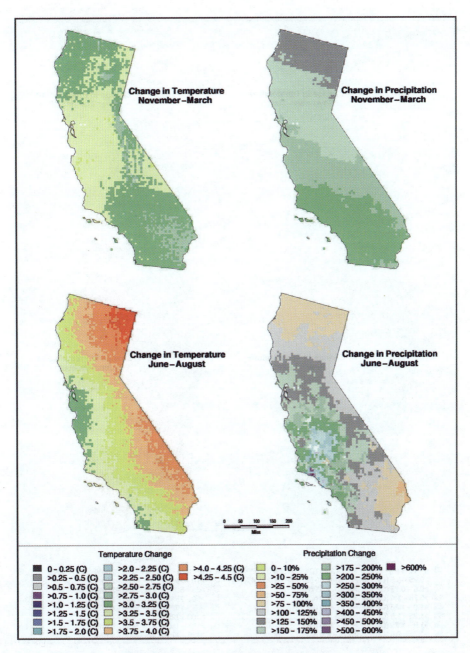

Change in Temperature
November–March

Change in Precipitation
November–March

Change in Temperature
June–August

Change in Precipitation
June–August

Temperature Change

■ 0 – 0.25 (C)	□ >2.0 – 2.25 (C)
■ >0.25 – 0.5 (C)	□ >2.25 – 2.50 (C)
■ >0.5 – 0.75 (C)	□ >2.50 – 2.75 (C)
■ >0.75 – 1.0 (C)	□ >2.75 – 3.0 (C)
■ >1.0 – 1.25 (C)	□ >3.0 – 3.25 (C)
■ >1.25 – 1.5 (C)	□ >3.25 – 3.5 (C)
■ >1.5 – 1.75 (C)	□ >3.5 – 3.75 (C)
■ >1.75 – 2.0 (C)	■ >3.75 – 4.0 (C)
■ >4.0 – 4.25 (C)	
■ >4.25 – 4.5 (C)	

Precipitation Change

□ 0 – 10%	□ >175 – 200%
□ >10 – 25%	■ >200 – 250%
■ >25 – 50%	□ >250 – 300%
□ >50 – 75%	□ >300 – 350%
□ >75 – 100%	□ >350 – 400%
□ >100 – 125%	□ >400 – 450%
■ >125 – 150%	□ >450 – 500%
□ >150 – 175%	■ >500 – 600%
	■ >600%

*Plate 1   Hadley changes in wet and dry seasonal precipitation and
temperature for 2080–99 compared to baseline*

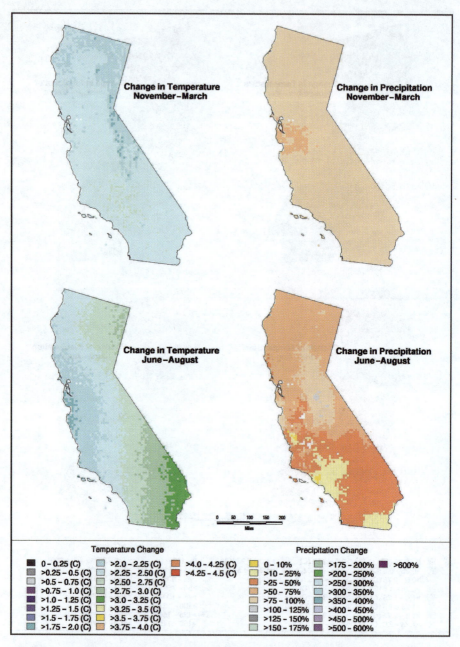

Change in Temperature
November–March

Change in Precipitation
November–March

Change in Temperature
June–August

Change in Precipitation
June–August

0   50   100   150   200
Miles

Temperature Change

■ 0 – 0.25 (C)	☐ >2.0 – 2.25 (C)
☐ >0.25 – 0.5 (C)	☐ >2.25 – 2.50 (C)
☐ >0.5 – 0.75 (C)	☐ >2.50 – 2.75 (C)
☐ >0.75 – 1.0 (C)	☐ >2.75 – 3.0 (C)
☐ >1.0 – 1.25 (C)	☐ >3.0 – 3.25 (C)
☐ >1.25 – 1.5 (C)	☐ >3.25 – 3.5 (C)
☐ >1.5 – 1.75 (C)	☐ >3.5 – 3.75 (C)
☐ >1.75 – 2.0 (C)	☐ >3.75 – 4.0 (C)

Precipitation Change

☐ 0 – 10%	☐ >175 – 200%	■ >600%
☐ >10 – 25%	☐ >200 – 250%	
☐ >25 – 50%	☐ >250 – 300%	
☐ >50 – 75%	☐ >300 – 350%	
☐ >75 – 100%	☐ >350 – 400%	
☐ >100 – 125%	☐ >400 – 450%	
☐ >125 – 150%	☐ >450 – 500%	
☐ >150 – 175%	☐ >500 – 600%	

☐ >4.0 – 4.25 (C)
☐ >4.25 – 4.5 (C)

Plate 2   PCM changes in wet and dry seasonal precipitation and temperature
for 2080–99 compared to baseline

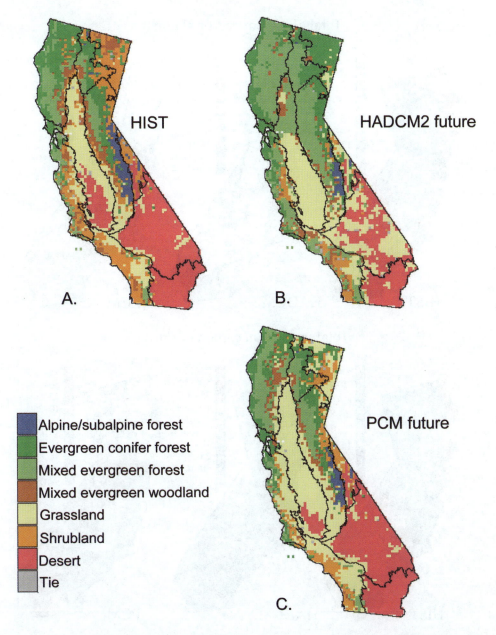

HIST

HADCM2 future

PCM future

Alpine/subalpine forest
Evergreen conifer forest
Mixed evergreen forest
Mixed evergreen woodland
Grassland
Shrubland
Desert
Tie

A.

B.

C.

*Plate 3   Distribution of the vegetation classes simulated (a) for the historical period
(1961–90) and for the future period (2070–99) of the (b) Hadley and (c)
PCM climate scenarios*

A.   Total ecosytem carbon (kg/m²)

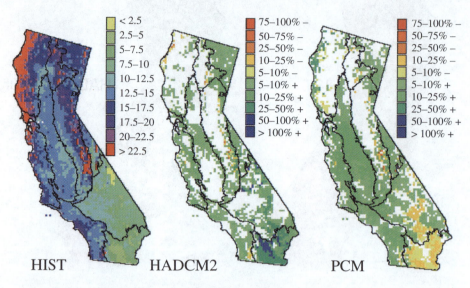

HIST   HADCM2   PCM

B.   Total vegetation carbon (kg/m²)

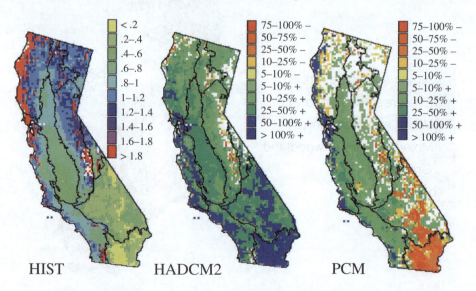

HIST   HADCM2   PCM

Plate 4   Distribution of (a) average total ecosystem carbon and (b) average annual total vegetation carbon for the historical period (1961–90) and for simulated changes in (2070–99) due to predicted Hadley and PCM climate changes

A.   Fire rotation period (years)

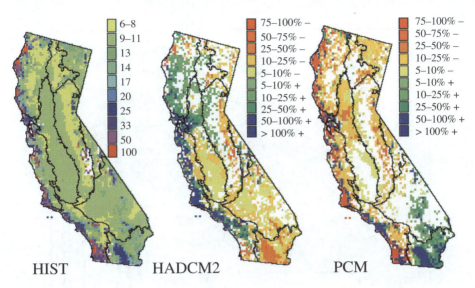

B.   Fire-line intensity (Btu ft⁻¹ s⁻¹)

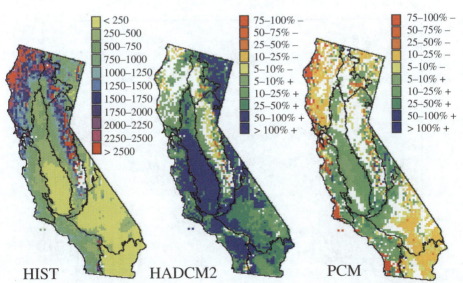

Plate 5   The distribution of the (a) fire rotation period and (b) average fire-line intensity per event for the historical period (1895–1994) and for simulated changes in (2000–2099) due to predicted Hadley and PCM climate changes

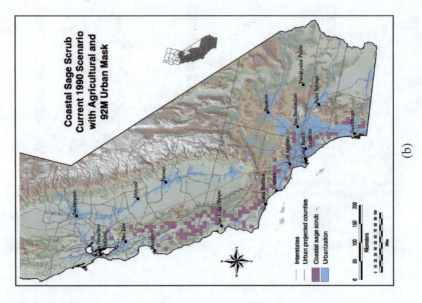

Coastal Sage Scrub
Current 1990 Scenario
with Agricultural and
92M Urban Mask

Interstates
Urban projected counties
Coastal sage scrub
Urbanization

0   50   100   150   200
Kilometers
0 10 20 30 40 50 60 70 80 90 100
Miles

(b)

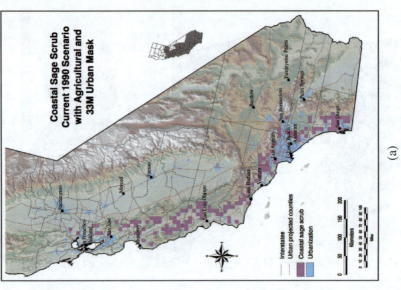

Coastal Sage Scrub
Current 1990 Scenario
with Agricultural and
33M Urban Mask

Interstates
Urban projected counties
Coastal sage scrub
Urbanization

0   50   100   150   200
Kilometers
0 10 20 30 40 50 60 70 80 90 100
Miles

(a)

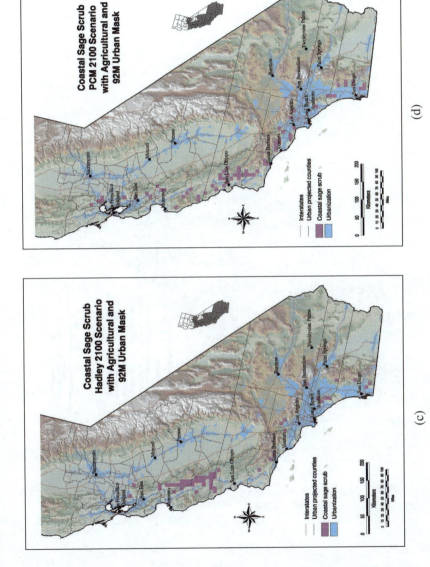

Plate 6 Potential distribution of CSS (a) with current conditions, (b) under 92 million population, (c) under the Hadley climate change scenario and 92 million population, (d) under the PCM climate change scenario and 92 million population

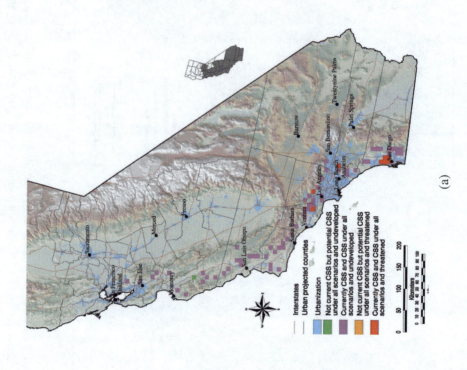

Interstates
Urban projected counties
Urbanization
Not current CSS but potential CSS under all scenarios and undeveloped
Currently CSS and CSS under all scenarios and undeveloped
Not current CSS but potential CSS under all scenarios and threatened
Currently CSS and CSS under all scenarios and threatened

Sacramento
San Francisco
Oakland
San Jose
Monterey
Merced
Fresno
San Luis Obispo
Santa Barbara
Ventura
Los Angeles
Long Beach
Anaheim
Barstow
San Bernardino
Palm Springs
Twentynine Palms
San Diego

Kilometers
0 10 20 30 40 50 60 70 80 90 100
0   50   100   150   200

(a)

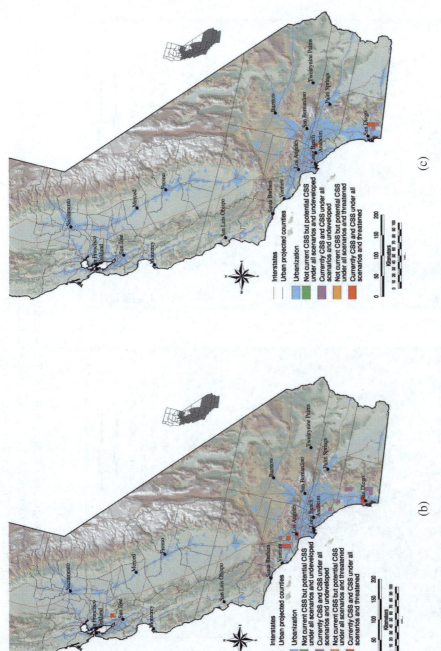

(c)

(b)

Plate 7   Potential distribution of CSS by (a) 2020 under all climate change scenarios and 45.5 million population, (b) 2060 under all climate change scenarios and 67 million population, (c) 2100 under all climate change scenarios and 92 million population

Interstates
Urban projected counties
Urbanization
Not current CSS but potential CSS under all scenarios and undeveloped
Currently CSS and CSS under all scenarios and undeveloped
Not current CSS but potential CSS under all scenarios and threatened
Currently CSS and CSS under all scenarios and threatened

Kilometers
0 10 20 30 40 50 60 70 80 90 100
0   50   100   150   200

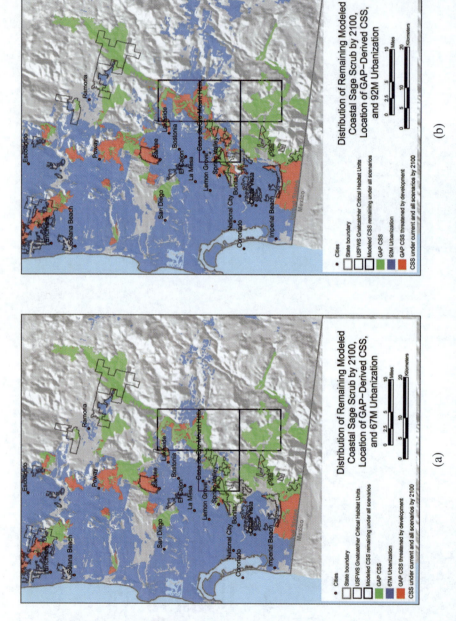

Plate 8   *Current distribution of GAP-derived CSS (1 ha resolution) in San Diego region; with estimated distribution of (a) 67 million urbanization and (b) 92 million urbanization. Figure also indicates areas designated as critical habitat units for the California gnatcatcher by the USFWS (2000)*

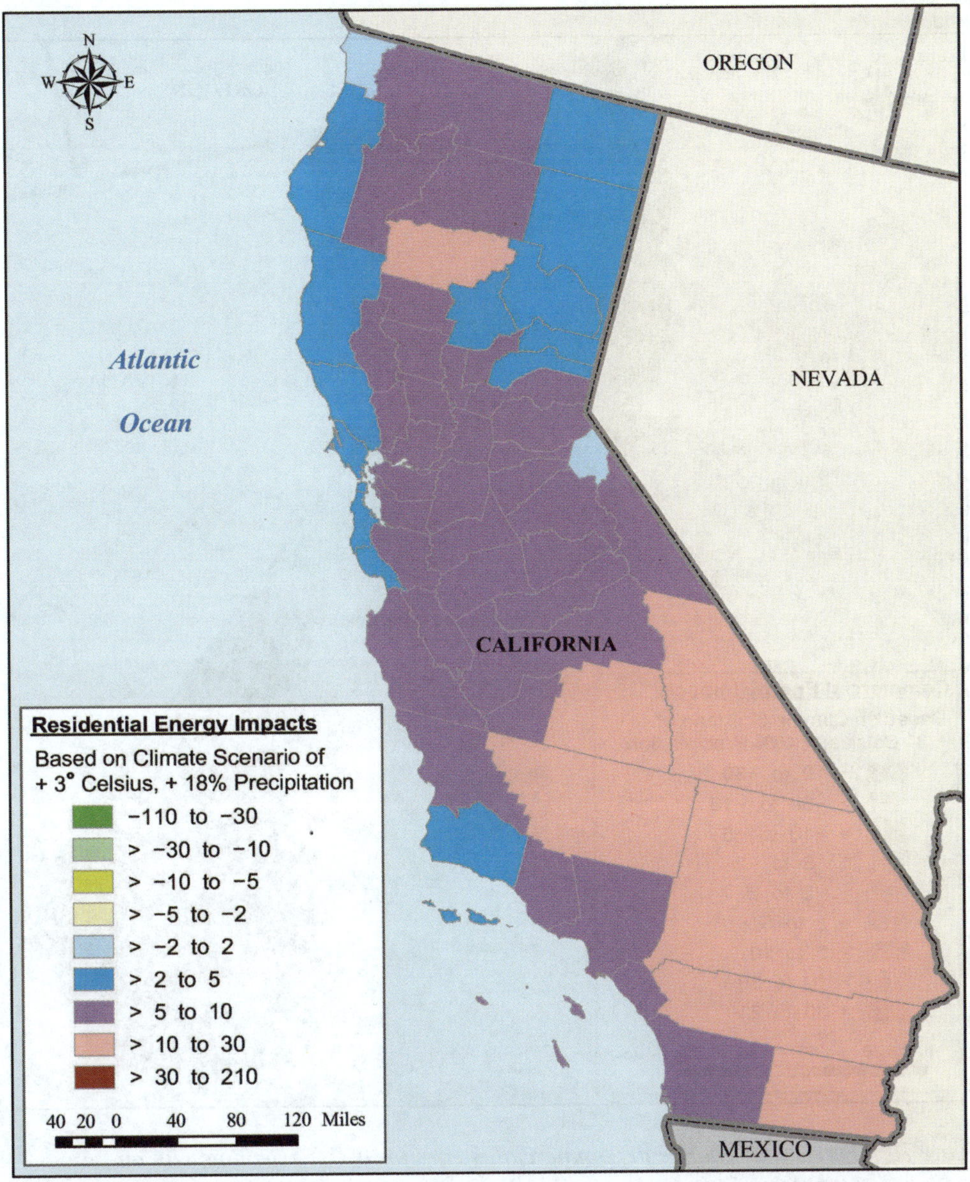

N
S
W
E
OREGON
NEVADA
Atlantic
Ocean
CALIFORNIA
MEXICO
Residential Energy Impacts
Based on Climate Scenario of
+ 3° Celsius, + 18% Precipitation
−110 to −30
> −30 to −10
> −10 to −5
> −5 to −2
> −2 to 2
> 2 to 5
> 5 to 10
> 10 to 30
> 30 to 210
40 20 0 40 80 120 Miles

Plate 9 Percentage change in residential energy for a 3°C warming with an 18 percent increase in precipitation

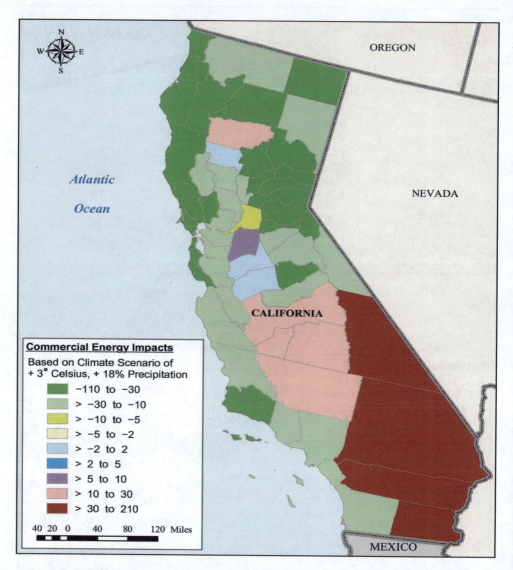

Plate 10   *Percentage change in commercial energy for a 3°C warming with an*
            18 percent increase in precipitation

temperatures and losses of carbon stocks to fire, resulted in a net decline in total ecosystem productivity during these relatively dry periods.

The simulated annual area of the state burned increased under all four incremental scenarios. Interestingly, fire increased under both the relatively wetter and relatively drier scenarios. The overall trend in the increase showed a strong relationship to the trend in increasing temperature. Peaks along the trend line are associated with relatively low annual precipitation and relatively high vegetation productivity (that is, available fuel) in preceding years. The relationship between area burned and vegetation productivity is also demonstrated by the slightly higher levels of total annual area burned under the wetter scenario at each level of temperature increase. Vegetation productivity was somewhat greater under each wetter scenario.

5.3.2.3 Future carbon budget

Simulated total ecosystem carbon for the entire state increases at a fairly steady and nearly equal rate under both future climate scenarios (Figure 5.3a). An increase of about 6 percent over the size of the historical pool is simulated at the end of the twenty-first century under both GCM scenarios[4] (Table 5.2). The rate and size of the increase in total ecosystem carbon under each scenario is largely driven by the simulated future trends in the soil and litter carbon pool (Figure 5.3b) which comprises over 90 percent of the total carbon pool (Table 5.2). The total live vegetation pool (Figure 5.3c) also shows a similar rate of increase under both scenarios throughout much of the future period. However, this similarity is the net result of different responses to the two scenarios by the wood and grass components of the total vegetation carbon pool. The wood component (Figure 5.3d) is the primary contributor to the increase in total vegetation carbon under the wetter Hadley scenario. The grass component is a greater contributor to the increase under the drier PCM scenario (Figure 5.3e).

The common response under both climate scenarios was an increase in statewide carbon storage produced by temperature-driven increases in productivity. Fluctuations in vegetation carbon storage (Figure 5.3c) showed a marked correspondence to fluctuations in future total annual precipitation under each scenario. However, the total amount of vegetation carbon storage was comparable under each scenario throughout much of the future simulation period, despite the disparate trends in precipitation. Up until the last few decades of the simulation period, the simulated response to the increasingly wetter Hadley (or increasingly drier PCM) scenario was an increased proportion of the total vegetation carbon stored as wood (or grass) carbon (Figures 5.3d and 5.3e). A precipitation effect on the trend of total vegetation carbon pool is clearly evident only near the end of the future period, when the two scenarios are most distinct in terms of total

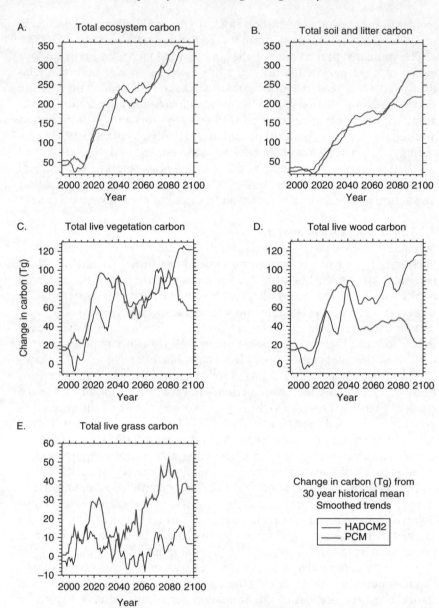

Figure 5.3   Smoothed percentage changes in (2070–99) relative to (1961–90) for storage in different carbon pools (a–e) under the Hadley and PCM scenarios

Table 5.2   Size of the historical carbon pools simulated for the state of
California and future changes in size simulated under the
Hadley and PCM climate scenarios[a]

Carbon pool	HIST (Tg)	Hadley change (Tg)	PCM change (Tg)
Total ecosystem	5765	+312	+325
Soil and litter	5305	+203	+246
Total live vegetation	461	+107	+78
Live wood	300	+99	+38
Live grass	163	+9	+41

Note: a. HIST values in teragrams are the mean weights for the 30 year (1961–90) base
period. Hadley and PCM change values in teragrams are the mean weights for the 30 year
(2070–99) future period subtracted from the mean weights for the historical period.

annual precipitation. At about 2085, the live vegetation carbon pool
declined under the drier PCM scenario but continued to increase under the
wetter Hadley scenario (Figure 5.3c). The decline in live vegetation carbon
under the PCM scenario is the net result of declines in both the wood
(Figure 5.3d) and grass (Figure 5.3e) carbon pools. Despite this decline in
vegetation carbon, total ecosystem carbon under the PCM scenario
(Figure 5.3a) was relatively unaffected because of a compensating increase
in soil and litter carbon (Figure 5.3b). The increase in the latter is apparently a lagged effect of the relatively steep increase in grass carbon just
before the period of vegetation carbon decline under the PCM scenario
(Figure 5.3e). Grass carbon is an especially effective contributor to soil
carbon in the model because of the fast turnover rate and because a large
proportion of grass biomass is below ground and protected from consumption by fire.

5.4   CONCLUSIONS

The MC1 simulation for the historical climate of California appeared to
achieve a reliable level of prediction, although validating a broad-scale
model of potential vegetation presents numerous difficulties. The simulation
of the coverage and distribution of vegetation classes was largely in agreement with current vegetation. The model also appeared to simulate accurately the relative distribution of tree and grass carbon, and average values
of the simulated total vegetation carbon per vegetation class were largely in
agreement with published values for equivalent classes. The simulated
mean fire return intervals fell within the range of independent estimates of

fire frequency in different vegetation types, and simulated fire intensities conformed to those expected for severe fire in different vegetation classes. Most of the severe fire years simulated by the model were coincident with observed drought years, and with observed severe fire years in the historical record for US Forest Service land in California.

More rigorous validation of the simulation results for the historical climate was difficult given the nature of the simulations. The model simulates dynamic ecosystem properties only as a function of climate and soils, and does not include the effects of land use practices on life-form mixtures, carbon stocks, and fire regimes. Also, the model does not include lags in vegetation change caused by migration and dispersal over a landscape fragmented by land use practices. Ongoing efforts to incorporate land use effects in MC1 and to increase the spatial resolution of the model simulations should increase the realism of the model results and facilitate their validation against observed data from the extant landscape.

The results of the MC1 simulations for California demonstrate certain ecosystem sensitivities and interactions that are likely to be features of the response of both natural and semi-natural (for example managed forest and rangeland) systems to the relatively certain rise in temperature and the less certain changes in precipitation. The most widespread response to the increase in temperature under both scenarios was a shift from conifer-dominated forests to mixed forests of conifers and evergreen hardwoods, primarily in the mountainous areas of the northern half of the state. Warmer temperatures increased the competitiveness of the evergreen hardwoods, which are less tolerant of low winter temperatures than conifers (Woodward, 1987). Higher temperatures also increased tree productivity in areas along the north-central coast and at high elevations where water stress is relatively low and growing season warmth is a constraint on growth. However, increased forest productivity with increased growing season temperature along the north-central coast would be contingent on the persistence of summer fog, which provides more than 30 percent of the annual soil moisture (Dawson, 1998). If increased temperatures were accompanied by a major decrease in coastal fog, the increased moisture stress would result in a decline in productivity or even the elimination of coast redwood forest and associated maritime forest types.

The simulated responses to changes in precipitation under the two future climate scenarios were more complex, involving not only a direct effect on vegetation productivity associated with changes in available soil moisture, but also changes in tree–grass competition that were mediated by fire. The persistence of a Mediterranean climate with dry summers was a key feature of the modeled response. Under the Hadley scenario, increased winter precipitation favored the more deeply rooted trees overall. More winter precipitation (together with CO_2-induced increases in water use efficiency) produced increases in tree carbon density sufficient to replace shrubland and woodland with forest in relatively mesic areas of the state. But in the more semi-arid regions, where annual precipitation continued to support co-dominance of trees and grasses, increased fire promoted by greater biomass and the persistently dry summers favored grasses, which recovered from fire-induced reductions in biomass more rapidly than trees. Consequently, under the Hadley scenario, increases were seen in both the forest cover at the expense of woodland and shrubland, and the productivity of grasslands.

Declines in winter precipitation under the PCM scenario produced a very different modeled response. A widespread decrease in tree carbon density increased the light, moisture, and nutrients available to the more drought-resistant grasses. The decrease in tree carbon density was not sufficient to convert much forested area to woodland or shrubland. But in the more ecotonal areas, increased grass biomass resulting from reduced competition with trees promoted more fire during the persistently dry summers, converting shrubland and woodland to grassland. Consequently, under the PCM scenario, there was an increase in the coverage of grassland at the expense of woodland and shrubland and a decrease in the productivity of forests.

Fire was a critical element in the simulated response to both future climate scenarios. The summer months were warmer and persistently dry under both scenarios, so differences in the modeled fire behavior and effects were primarily a response to differences in simulated fuels. The modeled extent of fire was most sensitive to changes in grass biomass. Changes in grass biomass produced changes in fuel loading and fuel bed structure that are strong determinants of the rate of fire spread simulated by the model. Increases in grass biomass were projected for different regions of the state under the Hadley and PCM scenarios, so the regions of simulated increases in fire area were also distinct. The simulated intensity of fire is more sensitive to the total amount of vegetation (that is, tree and grass) carbon available for consumption. Different ratios of tree and grass carbon produced similar levels of total vegetation carbon under the two scenarios, resulting in a greater degree of overlap for regions of increased fire intensity than for regions of increased fire area.

Although none of the model simulations for the different climate scenarios should be taken as predictions of the future, it is evident from the results that all the natural ecosystems of California, whether managed or unmanaged, are likely to be affected by changes in climate. Changes in temperature and precipitation will alter the structure, composition, and productivity of vegetation communities, and wildfires may become more frequent and intense. The incidence of pest outbreaks in forests stressed by a changing climate could act as a positive feedback on the frequency and intensity of fire. Non-native species that are pre-adapted to disturbance could colonize altered sites in advance of native species, preventing the already problematical redistribution of natives across a landscape highly fragmented by land use practices. Both plants and animals already stressed by human development will be further stressed by climate change. Some may not be able to adapt, and the number of threatened and endangered species could rise significantly. Tree species better adapted to a changed climate could be planted in forests managed for wood production, but better adapted species may not have the same market value (for example conifer species versus hardwood species). The expansion of grasslands under a drier climate might benefit grazing livestock, but any gains might be offset by decreased water availability.

Considerable uncertainty exists with respect to regional-scale impacts of global warming. Much of this uncertainty resides in the differences among the different climate scenarios as illustrated in this study. In addition, models that translate climatic scenarios into projections of ecosystem impacts can always be improved through re-examination and improvement of model processes. Nevertheless, the results of this study underscore the potentially large impact of climate change on California ecosystems, and the need for further use and development of dynamic vegetation models using various ensembles of climate change scenarios.

ACKNOWLEDGMENTS

We would like to thank the Electric Power Research Institute and the California Energy Commission for project support. We would also like to thank Christopher Daly of the Spatial Climate Analysis Service at Oregon State University for supplying spatially distributed, historical climate data, and Aiguo Dai of NCAR's Climate Analysis Section for providing the PCM future climate scenario.

NOTES

1. More details are provided in Appendix IV of EPRI's report to the California Energy Commission available at http://www.energy.ca.gov/reports/2003-10-31_500-03-058CF_ A04.PDF.
2. Figures comparing the distribution of vegetation under current climate as simulated by MC1 and maps of current vegetation distribution are in http://www.energy.ca.gov/ reports/2003-10-31_500-03-058CF_A04.PDF.

3. Figures showing trends in NBP and annual area burned as simulated by MC1 for the incremental scenarios are in http://www.energy.ca.gov/reports/2003-10-31_500-03-058CF_A04.PDF.
4. Carbon budgets were not calculated for the incremental scenarios.

REFERENCES

Aber, J., R.P. Neilson, S. McNulty, J.M. Lenihan, D. Bachelet, and R.J. Drapek. 2001. Forest processes and global environmental change: Predicting the effects of individual and multiple stressors. *Bioscience* **51**(9): 735–51.
Atjay, G., P. Ketner, and P. Duvigneaud. 1979. Terrestrial primary production and phytomass. In *The Global Carbon Cycle*, B. Bolin, E. Degens, S. Kempe, and P. Ketner (eds). John Wiley & Sons, New York, pp. 129–82.
Bachelet, D., J. Lenihan, C. Daly, and R. Neilson. 2000. Interactions between fire, grazing and climate change at Wind Cave National Park, SD. *Ecological Modelling* **134**: 219–24.
Bachelet, D., J. Lenihan, C. Daly, R. Neilson, D. Ojima, and W. Parton. 2001b. *MC1: A Dynamic Vegetation Model for Estimating the Distribution of Vegetation and Associated Ecosystem Fluxes of Carbon, Nutrients, and Water*. General Technical Report PNW-GTR-508. USDA Forest Service, Pacific Northwest Research Station, Portland, OR.
Bachelet, D., R.P. Neilson, J.M. Lenihan, and R.J. Drapek. 2001a. Climate change effects on vegetation distribution and carbon budget in the US. *Ecosystems* **4**: 164–85.
Barbour, M., B. Pavlik, F. Drysdale, and S. Lindstrom. 1993. *California's Changing Landscapes: Diversity and Conservation of California Vegetation*. California Native Plant Society, Sacramento.
California Division of Tourism. 2001. Visitor statistics. Available at http://gocalif.ca.gov/research/visitor.html.
Daly, C., D. Bachelet, J. Lenihan, W. Parton, R. Neilson, and D. Ojima. 2000. Dynamic simulations of tree-grass interactions for global change studies. *Ecological Applications* **10**: 449–69.
D'Antonio, C.M. and P.M. Vitousek. 1992. Biological invasions by exotic grasses, the grass/fire cycle, and global change. *Annual Review of Ecology and Systematics* **23**: 63–87.
Davis, F.W., D.M. Stoms, A.D. Hollander, K.A. Thomas, P.A. Stine, D. Odion, M.I. Borchert, J.H. Thorne, M.V. Gray, R.E. Walker, K. Warner, and J. Graae. 1998. *The California Gap Analysis Project*, Final Report. University of California, Santa Barbara, CA.
Dawson, T. 1998. Fog in the California redwood forest: Ecosystem inputs and use by plants. *Oecologia* **117**: 476–85.
Field, C., G. Daily, F. Davis, S. Gaines, P. Matson, J. Melack, and N. Miller. 1999. *Confronting Change in California: Ecological Impacts on the Golden State. Report of the Union of Concerned Scientists and the Ecological Society of America*. UCS Publications, Cambridge, MA.
Gutowski, W.J., Z. Pan, C.J. Anderson, R.W. Arritt, F. Otieno, E.S. Takle, J.H. Christensen, and O.B. Christensen. 2000. *What RCM Data are Available for California Impacts Modeling?*, California Energy Commission Workshop on

Climate Change Scenarios for California, California Energy Commission, Sacramento, 12–13 June.

Heinselman, M. 1973. Fire in the virgin forests of the boundary waters canoe area, Minnesota. *Quaternary Research* 3: 329–82.

Hickman, J.C. 1993. *The Jepson Manual: Higher Plants of California*. University of California Press, Berkeley, CA.

Holland, V. and D. Keil. 1995. *California Vegetation*. Kendall/Hunt Publishing Company, Dubuque, IA.

Houghton, R. and J. Hackler. 2000. Changes in terrestrial carbon storage in the United States. 1: The roles of agriculture and forestry. *Global Ecology and Biogeography* 9: 125–44.

Husari, S. and K. McKelvey. 1996. Fire-management policies and programs. In *Sierra Nevada Ecosystem Project: Final Report to Congress. Vol. II: Assessments and Scientific Basis for Management Options*, F. Davis (ed.). University of California Centers for Water and Wildland Resources Report 36, Davis, pp. 1101–17.

Jensen, D., M. Torn, and J. Harte. 1993. In *Our Own Hands: A Strategy for Conserving California's Biological Diversity*. University of California Press, Berkeley, CA.

Karl, T.R. 1986. The sensitivity of the Palmer Drought Severity Index and Palmer's Z-Index to their calibration coefficients including potential evapotranspiration. *Journal of Climate and Applied Meteorology* 25: 77–86.

Keane, R., C. Hardy, and K. Ryan. 1997. Simulating effects of fire on gaseous emissions and atmospheric carbon fluxes from coniferous forest landscapes. *World Resource Review* 9(2): 177–205.

Lenihan, J.M., C. Daly, D. Bachelet, and R.P. Neilson. 1998. Simulating broad-scale fire severity in a dynamic global vegetation model. *Northwest Science* 72: 91–103.

Lenihan, J.M., R. Drapek, D. Bachelet, and R.P. Neilson. 2003. Climate change effects on vegetation distribution, carbon, and fire in California. *Ecological Applications* 13(6): 1667–81.

Lertzman, K., J. Fall, and B. Dorner. 1998. Three kinds of heterogeneity in fire regimes: At the crossroads of fire history and landscape ecology. *Northwest Science* 72 (Special Issue): 4–23.

Martin, R. and D. Sapsis. 1995. A synopsis of large or disastrous wildlands fires. In *The Biswell Symposium: Fire Issues and Solutions in Urban Interface and Wildland Ecosystems*, D. Weise and R. Martin (technical coordinators). General Technical Report PSW-GTR-158. USDA Forest Service Pacific Southwest Research Station, Albany, CA, pp. 15–17.

McKelvey, K. and K. Busse. 1996. Twentieth-century fire patterns on Forest Service lands. In *Sierra Nevada Ecosystem Project: Final Report to Congress. Volume II: Assessments and Scientific Basis for Management Options*, F. Davis (ed.). University of California Centers for Water and Wildland Resources Report 36, Davis, pp. 1119–38.

National Assessment Synthesis Team (eds). 2001. *Climate Change Impacts on the United States: Foundation Report*. US Global Change Research Program. Cambridge University Press, New York.

Neilson, R. 1995. A model for predicting continental-scale vegetation distribution and water balance. *Ecological Applications* 5(2): 362–85.

Parton, W., D. Schimel, D. Ojima, and C. Cole. 1994. A general study model for soil organic model dynamics, sensitivity to litter chemistry, texture, and management. SSSA Special Publication 39. *Soil Science Society of America* 147–67.

Peterson, D. and K. Ryan. 1986. Modeling postfire conifer mortality for long-range planning. *Environmental Management* **10**: 797–808.

Rothermel, R. 1972. A Mathematical Model for Fire Spread Predictions in Wildland Fuels. USDA Forest Service Research Paper INT-115. USDA Forest Service, Intermountain Forest and Range Experiment Station, Ogden, UT.

Rothermel, R. 1983. How to Predict the Spread and Intensity of Forest and Range Fires. General Technical Report INT-143. USDA Forest Service, Intermountain Forest and Range Experiment Station, Ogden, UT.

Scheffer, M., S. Carpenter, J. Foley, C. Folke, and B. Walker. 2001. Catastrophic shifts in ecosystems. *Nature* **413**: 591–6.

Skinner, C. and C. Chang. 1996. Fire regimes, past and present. In *Sierra Nevada Ecosystem Project: Final Report to Congress. Vol. II: Assessments and Scientific Basis for Management Options*, F. Davis (ed.). University of California Centers for Water and Wildland Resources Report 36, Davis, CA, pp. 1041–69.

Strauss, D., L. Bednar, and R. Mees. 1989. Do one percent of forest fires cause ninety-nine percent of the damage? *Forest Science* **35**: 319–28.

Swetnam, T. and J. Betancourt. 1998. Mesoscale disturbance and ecological response to decadal climatic variability in the American Southwest. *Journal of Climate* **11**: 3128–42.

Turner, M. and W. Romme. 1994. Landscape dynamics in crown fire ecosystems. *Landscape Ecology* **9**(1): 59–77.

US Fish and Wildlife Service. 2001. Number of listed species in each state or territory. Available at http://ecos.fws.gov/webpage/usmap.html?&status=listed.

van Wagner, C.E. 1993. Prediction of crown fire behavior in two stands of jack pine. *Canadian Journal of Forest Research* **23**: 442–9.

Woodward, F. 1987. *Climate and Plant Distribution*. Cambridge University Press, New York.

6. Biodiversity changes and adaptation

Hector Galbraith, Joel B. Smith, and Russell Jones

6.1 INTRODUCTION

California's great variability in elevation, topography, soils, and climate makes it perhaps the most ecologically diverse area of North America (Schoenherr, 1992; Field et al., 1999; Wilkinson, 2002). Unfortunately, many of its unique or rare ecological resources are already under considerable anthropogenic stress. Since the European colonization, human activities, particularly agriculture and urbanization, have resulted in widespread and severe habitat loss; for example, approximately 86 percent of the wetlands in the Central Valley have been destroyed since the 1850s (Frayer et al., 1989). With continued human population growth, the potential for further loss and modification of the remaining natural habitats is significant. Another important and growing anthropogenic stressor is the intentional or unintentional introduction of exotic species, which has the consequences of displacing competitors and disrupting ecosystem processes. Approximately 20 percent of the state's current flora are introduced species, and some particularly successful exotics dominate large and previously diverse areas; yellow star thistle (*Centaurea solstitialis*), for example, now infests more than 10 million acres (4 million hectares). Many of these stressors continue to increase in intensity and range, posing difficulties for conserving the remaining resources.

The advent of global climate change has introduced yet another stressor into this already deteriorating situation. Acting through the changes that it may cause to vegetation communities, climate change has the potential to be a substantial new disruptor of California's ecological landscape (see Chapter 5).

Identifying adaptation measures to address this new situation will require that we understand how climate change might affect the spatial extent and distribution of the state's major ecosystems and vegetation communities. Studies of the possible effects of climate change on vegetation communities are an essential first step toward this goal (for example

Bachelet et al., 2001; Lenihan et al., 2003). However, predictive studies of areas highly modified by past (non-climate) human activities that do not consider these other stressors do not paint a realistic picture of how future ecological landscapes may be changed. Climate change does not happen in a vacuum, but may be just one more stressor exerting its effects on an ecological landscape that is already challenged and altered by many other important stressors. In California, urbanization, agriculture, habitat destruction, introduction of exotic species, contaminants, and over-exploitation of ecological resources all have been major forcing factors that produced the current landscape, and many of these continue to exert changes (particularly urbanization, agriculture, and the introduction of exotic species). The effects of climate change may be separate from those of the existing stressors, or they may exacerbate or sometimes mitigate them. For example, Galbraith et al. (2002) showed that future losses of coastal wetlands in southern San Francisco Bay because of sea level rises resulting from climate change may be magnified by water extractions from the underground aquifer, causing the land surface to subside. In this case, focusing on climate change alone leads to seriously underestimating potential future rates of habitat loss. Thus, when predicting the future effects of climate change on ecological resources in California, it is important to take an approach in which the intersecting effects of all the important stressors are integrated.

The study in Chapter 5 projected major reductions in the extents of Mediterranean shrubland (chaparral), C_3 grasslands, conifer savanna, tundra, and boreal conifer forest under all climate change scenarios. Less marked, but still fairly consistent, reductions are projected for arid shrubland, mixed xeromorphic woodland, and continental temperate conifer forest. All scenarios result in major projected increases in the spatial extents of C_4 grassland and warm temperate mixed forest.

Urbanization modeling projects major increases in urbanized land in the state by 2020 and 2060 (25 percent and 69 percent increases, respectively), and at least a doubling in this area by 2100 under the high population-growth scenario. A doubling in the area of urbanized land in the state (already a major landscape feature) will have important implications for the actual expression of the potential vegetation changes projected by the study described in Chapter 5.

An integrated approach is vital if we hope to develop meaningful results that can be used in, for example, conservation planning and land management. This study is the first attempt that has been made to forge this new analytical method. In this study we use the climate change scenarios described in Chapter 4, potential vegetation change (Chapter 5), and urbanization levels (Chapter 3) to evaluate the potential overall effects on the status and distribution of California's major vegetation communities. Our overall objectives were as follows:

- To integrate the potential effects of these major stressors into a more comprehensive appraisal of how the current ecological landscapes may change between now and 2100.
- To evaluate the relative contributions of urbanization and climate change to the estimated ecological effects, and to determine the proportional impacts of urbanization and climate change on the distributions of the community types, how their interaction will change over time, and how they will vary among ecosystems and under different climate change scenarios.
- To use the results of these analyses to explore the utility of this approach in potential adaptation planning. This can be done by estimating where particularly valued vegetation communities may survive, but be threatened by urbanization, and could suggest areas for conservation.
- To identify what further work is needed to refine this approach, so that the ecological impacts of climate change can be evaluated.[1]

6.2   METHODS

Our overall approach in this analysis was to combine estimations of changes in the spatial extents and distributions of potential vegetation community types under future climate change scenarios (Chapter 5) with spatial projections of future urban development patterns (Chapter 3) on a geographic information system (GIS) platform (Figure 6.1). Both studies estimate changes until 2100. We then analyzed the combined data sets to identify and quantify future vegetation community changes that could result from climate change, urbanization, or the interaction of both. For finer resolution analyses of one community type, coastal sage scrub (CSS), we combined data from the California Gap Analysis Project (GAP) (Davis et al., 1998) with the urbanization and potential vegetation projections. Figure 6.1 charts our approach to this analysis.

The study area considered was not the entire state of California, but a more limited area extending from the Central Valley south to the border with Mexico (Figure 6.2). This was the area for which urbanization projections were available. The area includes counties that are currently heavily urbanized. It is possible that more northern California counties would experience such widespread urbanization as well, particularly under high population-growth scenarios.

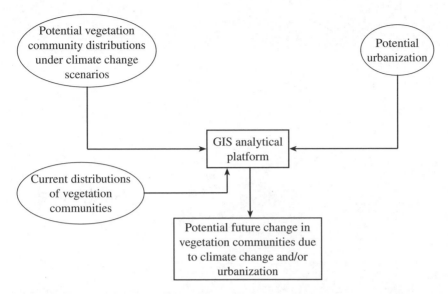

Figure 6.1   Methodological approach

We used the two general circulation model (GCM) scenarios (Hadley and parallel climate model, PCM) and the incremental scenarios described in Chapter 4.

Once combined on the GIS platform, the urbanization results were overlaid on the potential vegetation results and used as a 'mask' to identify and exclude projected potential vegetation changes that would be unlikely to occur because of current or future urbanization. The current distribution of agricultural lands was also included in this mask. However, the spatial extent of agricultural land in California will likely be stable or contract slightly in the next few decades (State Water Department, 1998; Wilkinson, 2002). Consequently, using the current distribution of agriculture as a 'future mask' is unlikely to affect seriously the future projections of vegetation community shifts.

Scale mismatches were noted between the two data sets from the studies in Chapters 3 and 5. Landis and Reilly in Chapter 3 estimated urbanization at a 1 ha scale and Lenihan and colleagues estimated vegetation distribution at a 100 km^2 (10 km grid cell size) scale. To make the data sets compatible, we scaled the urbanization projections up to 100 km^2 by assuming that grids 50 percent or more urbanized are urban and those less than 50 percent urbanized are natural or semi-natural vegetation communities. The GAP data have a spatial resolution of 1 ha and were therefore directly compatible with the urbanization projections. The decision to call

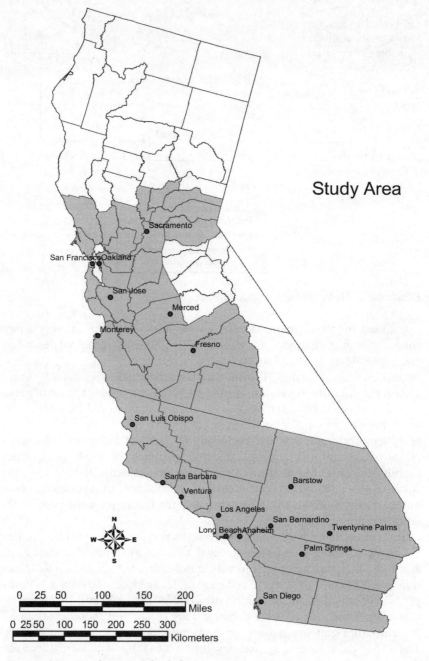

Figure 6.2   Study area (shaded counties)

cells that are 50.1 percent urbanized 'completely urban' is arbitrary. The degree to which any level of urbanization will be disruptive to ecosystems varies considerably. Some may be healthy despite higher levels of urbanization, but the health of others may be disrupted at much lower levels of urbanization.

6.2.1   Limitations

One of the uncertainties in MC1, the vegetation model used in this analysis, is that it estimates where vegetation should be present, based on which vegetation community type is best correlated with estimated climate in a particular grid box. The model does not estimate migration or factors that impede it. This adds uncertainty to estimates of change in the location of vegetation communities. Given the heterogeneity of ecosystems in California, however, species will not have to migrate very far, which increases the likelihood that vegetation types will migrate to the new locations estimated by the model. Another uncertainty is that the model does not estimate shifts among individual species. Even though the model may estimate the shift from one vegetative type to another within a grid cell, it does not estimate the composition of the changed vegetation. We can presume that at least some species in a vegetative community will be in grid boxes estimated to be dominated by that community type. We do not know if most or all species will be present. We examine CSS below, but keep in mind that the vegetation model estimated location of a class of vegetation that includes CSS, but not solely CSS.

One of the key uncertainties in the projection of urbanization changes is that we use only one development scenario. The study in Chapter 3 estimated development under different scenarios, but we used only the baseline scenario. This assumes that current patterns continue through the end of the twenty-first century. Indeed the patterns are based on the examination of ten recent years of data. Housing preferences, laws, and other factors that affect development could change over the century, in turn affecting the extent of land cover. With the population scenarios used in this study, new employment centers will arise, but it is difficult to know where these centers will spring up. Landis and Reilly (Chapter 3) used their best judgement. However, most likely development will proceed out from currently heavily developed areas, and in particular into the low-lying, near-urban areas in southern California (Chapter 3 provides more discussion of the model and results).

Even in light of these uncertainties, exercises such as this are important in helping us gain a better understanding of the behavior of the systems we are studying and their potential sensitivities to the forces that drive them. In addition, we can gain insight into the relative effects of climate change and urbanization, plausible future outcomes, and potentially important areas for conservation.

6.3   RESULTS

6.3.1   All Community Types

Except in areas where urban development is unlikely because of topography and climate (higher elevation tundra and boreal forest communities), habitat losses are projected for many communities as a result of urbanization alone (Table 6.1). Nevertheless, the greatest changes for all community types and losses for most community types occur when climate change is also included (Table 6.1). By 2100, climate change is projected to be responsible for much larger changes in habitat area for most communities than urban growth.

Under all climate change scenarios, tundra, boreal conifer forest, Mediterranean shrubland, C_3 grasslands, and subtropical arid shrubland are projected to decrease in extent (Table 6.1), most of them significantly. However, Table 6.1 also indicates that the extent of some communities is estimated to increase under climate change: C_4 grasslands and warm temperate and subtropical mixed forests benefit under all climate change scenarios, and maritime temperate conifer forest benefits under the relatively wet Hadley scenario.

Table 6.1   Projected effects (changes in acreage) of urbanization and climate change scenarios on vegetation types by 2100

Community type	Percent change resulting from urbanization	Percent change resulting from climate change
Tundra	0	$-42.9 \Leftrightarrow -82.1$
Boreal conifer forest	0	$-37.9 \Leftrightarrow -51.7$
C_3 grassland	-9.5	$0 \Leftrightarrow -84.8$
C_4 grassland	-2.0	$+236.6 \Leftrightarrow +510.9$
Mediterranean shrubland	-13.2	$-30.3 \Leftrightarrow -59.4$
Subtropical arid shrubland	-1.9	$-1.8 \Leftrightarrow -51.4$

6.3.2   Landscape Ecological Diversity

California's ecological diversity is apparent both within communities and
at the landscape level, where large numbers of distinctly different habitat
types occur in close proximity. We investigated the potential effects of
future urbanization and climate change on this landscape ecological diversity by calculating Shannon–Wiener indices (Krebs, 1978) from the current
potential vegetation data. We then compared the results with projected
indices for the future potential vegetation under climate change and urbanization. Shannon–Wiener indices are basically nonparametric scores of the
degree to which a landscape exhibits habitat heterogeneity. A highly diverse
landscape (that is, one in which many different habitat types are in close
proximity) would have a higher index than one under a monoculture or one
that has relatively few habitats.

Figure 6.3 displays the changes in landscape diversity as measured using
the Shannon–Wiener index based on the estimated changes in community
type extents. These data suggest that urbanization alone will result in a
slight reduction in diversity, but climate change and urbanization together
may result in greater impacts at the landscape level. The degree and direction of this effect vary with the temperature and precipitation change
assumptions – little or no reduction is estimated under the relatively wet

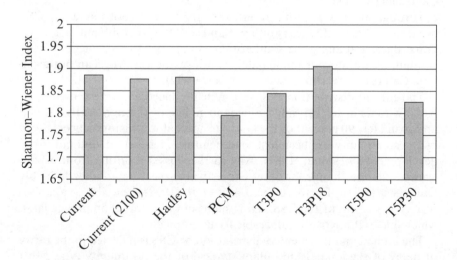

Figure 6.3   Landscape diversity as measured using the Shannon–Wiener
index based on the estimated changes in community type
extents under climate change scenarios by 2100 and 92 million
estimated population

Hadley or T3P18 scenarios. Reductions are projected under all the other scenarios, however, with the greatest reductions under the relatively dry PCM and T5P0 scenarios because of the increased aridity that is contingent on these scenarios and its constraining effect on the potential for plant community development. Under hotter and drier conditions, grasslands move into areas currently occupied by shrubs and woodlands and diversity of arid areas remains unchanged (J. Lenihan, Oregon State University, Forestry Science Laboratory, personal communication, 11 October 2002).

6.3.3   California Coastal Sage Scrub

Mediterranean shrubland was one of the community types projected to be most affected by urbanization and climate change. Table 6.1 projects decreases in spatial extent of between 30 percent and 60 percent of its current area. Mediterranean shrubland comprises a number of more or less different community types. One of these is CSS, which is a shrub community dominated typically by drought-deciduous species of the genera *Artemisia* (sageworts), *Eriogonum* (buckwheats), and *Salvia* (sages). It occurs mainly closer to the coast, on sandier soils, and in situations that are more xeric than those where other chaparrals are found (Keeley and Keeley, 1988). CSS currently extends from the central California coast south into northern Baja California (Mexico).

Throughout its California range, the largest contiguous areas of CSS occur on the Pismo Dunes (south of San Luis Obispo), on Vandenberg Air Force Base, and along the southern coast adjacent to San Diego. Larger contiguous areas occur immediately south of the US–Mexican border in Baja California (Westman and Malanson, 1992).

The current distribution of CSS is much reduced and fragmented compared with its historical range. Since the middle of the eighteenth century, approximately 90 percent of the land surface that was historically occupied by the community has been lost to agricultural, residential, and industrial development (Westman, 1981). Any further losses resulting from urbanization, climate change, or other causes will exacerbate an already highly tenuous situation. Furthermore, Davis et al. (1998) showed that approximately 70 percent of the CSS that remains in California is on private land, which makes it relatively vulnerable to development.

The limited and fragmented distribution of CSS is reflected in the rarity of many of the animals and plants typical of the community type. Most prominent of these has been the California gnatcatcher, *Polioptila californica*, which the US Fish and Wildlife Service (USFWS) listed in 1993 as threatened under the federal Endangered Species Act. Protecting this species and its supporting habitat has been the flashpoint for many of the coastal conservation–development conflicts in recent years in the Los Angeles and San Diego areas.

We focused more closely on CSS to investigate, with greater spatial resolution, the projected ecological effects of future urbanization and climate change. We then used this finer-grain analysis to evaluate the extent to which our overall analytical approach permits us to identify and evaluate potential conservation constraints and opportunities. We first combined the potential vegetation (Chapter 5) and the urbanization projections (Chapter 3) to investigate potential effects of urbanization and climate change on the future distribution of the CSS community at a 10 km grid cell scale. In the potential vegetation projections, CSS is a component of Mediterranean shrubland. To partition out the CSS distributions, we assumed that all Mediterranean shrubland within 70 km of the southern coast, at elevations of less than 700 m, and west of any topographic barrier, is CSS.

Table 6.2 shows the results of overlaying the future urbanization projections and climate scenarios on the modeled distribution of CSS. Under a population scenario of 92 million people statewide, urbanization alone is estimated to result in a more than 20 percent further reduction in the current acreage of CSS. Because most of the CSS habitat losses in the last few decades resulted from the expansion of residential areas, this is not unexpected. However, the estimated losses increase greatly when climate change is factored into the model – to between approximately 20 percent and 60 percent for 2060, and to between 30 percent and 63 percent for 2100 – roughly a doubling to tripling of the projected losses caused by urbanization alone. Interestingly, the greatest rate of loss occurs between the present day and 2020 in most of the urbanization plus climate change

Table 6.2   *Projected effects of future urbanization and climate change[a] on CSS area (percent changes in acres)*

Year	Urban. only	Urban. plus HadCM2	Urban. plus PCM	Urban. plus T3P0	Urban. plus T3P18	Urban. plus T5P0	Urban. plus T5P30
2020	−9.2	−0.7	−37.6	−19.9	−16.3	−19.9	−18.4
2060	−14.2	−39.7	−61.7	−22.0	−21.3	−24.1	−22.7
2100 (92 million population scenario)	−21.3	−63.1	−52.5	−34.0	−32.6	−34.0	−29.8

Note: a. Tables 4.2 and 4.3 in Chapter 4 give temperature and precipitation changes for each scenario.

scenarios. This may be because these rates of destruction leave little CSS to be further modified after 2050 in areas where development is feasible.

Plate 6a shows the distribution of CSS with current population. Plate 6b shows the potential distribution of CSS, assuming a human population of 92 million. Plates 6c and 6d show the projected 2100 distributions of CSS, also assuming a human population of 92 million and the Hadley and PCM scenarios. These figures show that, in general, the greatest losses of CSS to urbanization are estimated to occur in the southern part of the community's range, particularly in the Los Angeles Basin and the Ventura coastline.

To identify potential CSS conservation opportunities in this southern part of the community's range, we identified those 10 km^2 grid cells that are currently CSS and are projected to stay CSS under all the climate scenarios at the three levels of population (Plate 7). This does not account for all potential climate changes; different vegetation models could yield different results, but our analysis gives a preliminary indication of where CSS may be most likely to survive climate change. In addition, we examined cells that are not currently CSS to see if any may become CSS under all the climate change scenarios.

Many cells were estimated to support CSS under all of the climate change scenarios by 2020. By 2060, far fewer cells support CSS. We found that only five cells would be likely to have CSS under all of the climate change scenarios by 2100. One cell is near Irvine, and the other four are east of San Diego.

We next overlaid the GAP distribution in southern California with the urbanization projections for that area. Because both these data sets are on a 1 ha scale, we can investigate the potential effects of urbanization and climate change on the distribution of this community type at a much finer resolution. Also, this area is where many of the rare species that are characteristic of CSS are found (for example the California gnatcatcher). The results, then, are particularly relevant to protecting these resources.

Plate 8 shows the projections to 2100 of this more detailed analysis for the 67 million and 92 million population scenarios for the area east of San Diego. (We did not examine the 10 km grid cell near Irvine because GAP shows it currently has little CSS.) The blue areas are the projected distributions of urbanization that are not currently CSS. The red areas are currently CSS and predicted to be urbanized, and the green areas are currently CSS and not threatened by development.

The heavily outlined 10 km grid cells (Plate 8) are the squares that the previous analysis projected would survive climate change. Thus, even in some of those areas that may survive climate change, further losses of CSS are projected to occur because of development. This is particularly the case in the western, northern, and central parts of the outlined area. By 2100, little of the current distribution of CSS is expected to survive, and only in steeper, more mountainous areas in the south of the outlined area where development may not reach. These areas, and the areas that are projected to survive climate change but are threatened by future development, could most effectively be the focus of future conservation efforts for this plant community type and its associated animals.

We can see that not all CSS need necessarily be lost. However, even if these areas were completely protected from development, under climate change there would still be a 7.8–46.8 percent loss of CSS in California by 2100. Also, it is inevitable that the fragments that survived would be smaller, farther apart, and closer to developed areas than under current conditions. Smaller and more dispersed patches that are adjacent to developed areas could result in increased species extinction rates, reduced recolonization rates, increased human disturbance, and the enhanced potential for invasion by exotic species and non-native predators. Thus the projected changes could have important implications for whether the surviving patches would be functionally equivalent to CSS as it exists now.

6.4 CONCLUSIONS

This analysis led us to several important conclusions about the overlay of vegetation changes with urbanization: The analytical approach and the results demonstrate the potential for combining estimates of change in location of vegetation with estimates of change in urbanization. This is an important methodological advance for situations where stressors other than climate change may be important. The approach used can partition the relative effects of these important stressors and project their changing contributions over time. By combining urbanization and climate change data sets, we can preliminarily identify areas that could still support valuable vegetation communities under climate change, but would be threatened by development. This may allow us to focus future conservation efforts more effectively.

Climate change is estimated to have a much larger impact on the distribution and spatial extent of the major vegetation community types studied (for example Mediterranean shrubland, C_3 grassland) than urbanization. Habitat heterogeneity in California may be reduced if climate change results in lowered soil moisture or substantially higher temperatures (that is, approximately 5°C). Diversity is estimated to increase slightly if climate change results in wetter conditions and smaller increases in temperature (that is, 3°C or less).

Coastal sage scrub is particularly threatened by future urbanization (more than 20 percent loss of existing habitat by 2100). Furthermore, climate change could result in three times greater loss of coastal sage habitat. Potential refugia for coastal sage scrub have been preliminarily identified. These, and areas where urbanization but not climate change threatens CSS, could be the focus of future conservation efforts.

These results should be treated as approximations. The urbanization projections are based on a number of very specific assumptions and assume that current trends will continue. The vegetation model did not specifically model distribution of CSS. Nevertheless, this kind of analysis may be most useful for identifying where much more focused and detailed studies could take place.

NOTE

1. Details of this study are provided at http://www.energy.ca.gov/reports/2003-10-31_500-03-058CF_A05.PDF.

REFERENCES

Bachelet, D., R.P. Neilson, J.M. Lenihan, and R.J. Draypek. 2001. Climate change effects on vegetation distribution and carbon budget in the United States. *Ecosystems* **4**: 164–85.

Davis, F.W., D.M. Stoms, A.D. Hollander, K.A. Thomas, P.A. Stine, D. Odion, M.I. Borchert, J.H. Thorne, M.V. Gray, R.E. Walker, K. Warner, and J. Graae. 1998. *The California Gap Analysis Project – Final Report*. University of California, Santa Barbara.

Field, C.B., G.C. Daily, F.W. Davis, S. Gaines, P.A. Matson, J. Melack, and N.L. Miller. 1999. *Confronting Climate Change in California. Ecological Impacts on the Golden State*. Union of Concerned Scientists, Cambridge, MA, and The Ecological Society of America, Washington, DC.

Frayer, W.E., D.D. Peters, and H.R. Pywell. 1989. *Wetlands of the California Central Valley Status and Trends: 1939 to mid-1980's*. US Fish and Wildlife Service, Washington, DC.

Galbraith, H., R. Jones, R. Park, J. Clough, S. Herrod-Julius, B. Harrington, and G. Page. 2002. Global climate change and sea level rise: Potential losses of intertidal habitat for shorebirds. *Waterbirds* **25**: 173–83.

Keeley, J.E. and S.C. Keeley. 1988. Chaparral. In *North American Terrestrial Vegetation*, M.G. Barbour and W.D. Billings (eds). Cambridge University Press, Cambridge, UK, pp. 165–207.

Krebs, C.J. 1978. *Ecology: The Experimental Analysis of Distribution and Abundance*. Harper International, New York.

Lenihan, J.M., R. Drapek, D. Bachelet, and R.P. Neilson. 2003. Climate change effects on vegetation distribution, carbon, and fire in California. *Ecological Applications* **13**(6): 1667–81.

Schoenherr, A.A. 1992. *A Natural History of California.* University of California Press, Berkeley, CA.

State Water Department. 1998. California Water Plan. Bulletin 160–98. Sacramento, CA.

US Fish and Wildlife Service (USFWS). 2000. 1:48 000 Scale Digital GIS Data, Final Critical Habitat for the Coastal California Gnatcatcher. Received 21 March 2002, via e-mail from Tony McKinney, GIS coordinator, Carlsbad, CA.

Westman, W.E. 1981. Factors influencing the distribution of species of Californian coastal sage scrub. *Ecology* **62**: 439–55.

Westman, W.E. and G.P. Malanson. 1992. Effects of climate change on Mediterranean-type ecosystems in California and Baja California. In *Global Warming and Biological Diversity*, R.L. Peters and T.E. Lovejoy (eds). Yale University Press, New Haven, CT, pp. 258–76.

Wilkinson, R. 2002. *The Potential Consequences of Climate Variability and Change for California: The California Regional Assessment.* US Global Change Research Program. Washington, DC.

7. Timber impacts

Robert Mendelsohn

7.1 INTRODUCTION

This chapter examines two distinct ways that climate change might affect the timber sector. The first analysis begins with the predicted changes in California's forestland and forest productivity that a range of climate scenarios would cause using the ecological model described in Chapter 5. The analysis calculates how the ecological changes will affect the production of timber in the state. Multiplying these changes in timber supply by price provides an estimate of the welfare impact of within-state effects. The second analysis examines the consequences of price changes from global forest impacts on Californians. Climate change affects not only California timber production but also timber production in the rest of the world. If worldwide production increases, for example, timber prices will fall, which will hurt California's timber producers but help the state's timber consumers. An earlier study predicted how different global climate scenarios might affect global timber production and thus timber prices (Sohngen et al., 2002). The second analysis explores the impacts of those price changes on both California producers and consumers.

The timber sector is a relatively small component of the economy of California. The timber harvested in 2000 sold for $452 million. In contrast, the state economy was estimated to be $1.33 trillion (year 2000 USD; California Department of Finance, 2003). The harvests are also a small fraction of national production. California's harvests for sawtimber were about 1.7 billion board feet (Table G-27 in California Department of Finance, 2003), whereas US production was about 51 billion board feet (Howard, 2001). Altogether about 17 million hectares of land in the state are suitable for timber production (Table G-29 in California Department of Finance, 2003). Nationwide, there are over 500 million hectares of commercial forest.

Nonetheless, climate change is predicted to have far-reaching effects on forests. Ecological models suggest that climate change will shift the geographic distribution of tree species (Emanuel et al., 1985; Shugart et al., 1986; Solomon, 1986; Neilson and Marks, 1994) and alter productivity

(Melillo et al., 1993) throughout the world. Over the twenty-first century, the ecological model from Chapter 5 tracks forestland, fires, and productivity for each biome.

Starting with these ecological predictions of changes, the analysis predicts changes in timber supply. Because this study is limited to timber market impacts, a great deal of the southern portion of the state is not explored because it is already too hot and too dry to support commercial forests. Although climate change is predicted to increase the amount of hardwood forests, this too is not included in this study because most of the California hardwood forest is not suitable for timber production. The study focuses on the predicted changes in the area and the productivity of softwoods to predict changes in timber supply.

Recent studies of climate change impacts on timber reveal that it is important to capture the dynamic path of ecological impacts (Sohngen and Mendelsohn, 1998; Sohngen et al., 1999, 2002). It is not enough to compare current production with a future equilibrium production level in 2100 because such comparisons do not reveal what happens throughout the century. Given that climate change is gradually unfolding over this period, that the ecological system is gradually responding, and that markets are slowly adapting, the dynamics of this problem are important. This study makes an important advance over past efforts to capture impact dynamics, in that this economic analysis is the first to rely on dynamic ecological modeling. In past studies, economists assumed linear dynamic paths on the basis of long-term equilibrium changes across ecosystems. However, in this study, the ecosystem model described in Chapter 5 predicts the dynamic ecological changes for California, which are used to predict changes in timber supply over time.

This study also explores the critical issue of scale. Previous studies have tended to rely on large-scale modeling. Typically, entire regions are captured in broad ecological models, where the smallest unit of observation is actually a large and heterogeneous area. These ecological predictions are then aggregated to fit available economic data on large contiguous areas (Sohngen et al., 1999, 2002). In this study, a geographically detailed model is used to capture small-scale phenomena.

The analysis also examined the predicted land use changes described in Chapter 3. However, most of the predicted growth in urban and residential land use over the next century is at the expense of farmland, scrubland, and hardwood forests. Because they are either high in the mountains or far to the north, the softwood forests used for timber in California would remain largely unaffected by such land use growth over the next century.

The second analysis, in examining the impact of climate change on timber prices, is based on the fact that timber is traded across nations so that the price is determined at the global scale. Climate change is predicted to increase global timber supply (Binkley, 1988; Bowes and Sedjo, 1993; Joyce et al., 1995; Burton et al., 1998; Sohngen and Mendelsohn, 1998; Sohngen et al., 1999, 2002; McCarl et al., 2000; Gitay et al., 2001), leading to price declines. This analysis takes the predicted price changes from one global forest economic model for different climate scenarios (Sohngen et al., 2002). Given these price reductions, we calculate the losses to producers and the welfare gains to consumers. The price (second) analysis does not exactly correspond to the within-state (first) analysis because the climate models are slightly different and the ecological models are not the same. However, the two analyses are sufficiently similar to get a sense of their relative importance.

The impacts of climate change on the non-market values of forested ecosystems are not evaluated in this chapter. Some of the effects on biodiversity are explored in Chapter 6. Previous studies indicate that warming is also likely to hurt the ski industry but improve summer recreation (Loomis and Crespi, 1999; Mendelsohn and Markowski, 1999; Pendleton and Mendelsohn, 2000). These effects, however, have not been measured in this study.

7.2 METHODS

This study examines the outcomes of four uniform climate change scenarios and two general circulation models (GCMs) (see Chapter 4). The uniform changes are the 3°C and 5°C warming with no precipitation change, a 3°C change with an 18 percent increase in precipitation, and a 5°C change with a 30 percent increase in precipitation.

The ecological consequences of each of these scenarios are explored using the MC1 model described in Chapter 5. The ecological model offers two important improvements over past impact studies. First, the ecological model predicts dynamic changes. In the past, economists have inferred linear dynamic pathways from equilibrium ecological results Although this crude linear approach yields a sense of ecosystem dynamics, the dynamic predictions of an ecological model are highly preferable.

Second, this ecological model calculates geographically detailed predictions. In the past, ecological predictions have been aggregated over large areas to match available economic data. For example, some US and global studies of timber used a single observation for the entire state of California (Sohngen and Mendelsohn, 1998; Sohngen et al., 1999, 2002). In this study, the geographically detailed ecological predictions are used to make county-specific economic analyses in California.

The terrestrial biosphere model (MC1) uses detailed climate predictions to estimate dynamic changes in the distribution of biomes and the productivity of timber species across the state. Of course, this is a difficult undertaking because ecology has yet to determine precisely how ecosystems might change over time. It is not clear how long current stands might survive and how quickly new stands of new species might get established.

Biomes are ecological types that represent accumulations of different species. In this study, we focus on the forested biomes that support softwoods. We selected softwoods as the focus of this study because the bulk of California's timber supply comes from softwoods. Almost 94 percent of harvests came from softwoods in 1996 (California Department of Finance, 2003). Further, most of the forests in California are softwoods, 12.9 million acres out of 16.7 million acres of forests (77 percent in 1996) (California Department of Finance, 2003).

Three biome types can support softwood timber species. The maritime temperate coniferous forest along the coast supports Douglas fir, western hemlock, Sitka spruce, western red cedar, (interior) lodgepole pine, Bishop pine, Monterey cypress, Monterey pine, and coastal redwoods. The continental temperate coniferous forest in the northern interior and lower Sierras supports Douglas fir, other true firs, ponderosa pine, Jeffrey pine, and (coastal) lodgepole pine. The boreal conifer forest along the high Sierras supports lodgepole pine, mountain hemlock, and other mountain pines.

MC1 predicts how much land is in each biome during each decade. This information is used to calculate the area of land that can be regenerated to timber. If the area is shrinking (growing), we assume that the regenerated land must also shrink (grow) proportionately. We do not assume, however, that the standing forest would change. There is reason to believe that the standing forest would not change as quickly as the model predicts for each decade. At the detailed geographical scale employed in this analysis, the ecological model predicts fairly large swings in territory between one biome and another during each decade. By altering planting only in response to these predictions, we may understate the full extent of the change. However, the model does not explicitly reflect the structure of a standing forest. In practice, long-lived species would not survive if thousands of hectares shifted in and out of each biome during each decade. By adjusting only regeneration, we can smooth these predicted rapid fluctuations of biomes in the model.

MC1 provides two alternative measures of productivity; it calculates net primary productivity (NPP) and change in forest carbon. The carbon measure reflects not only growth but also fire. We compute changes in growth rates on the basis of changes in NPP or carbon accumulation and compare the results. The changes of productivity $\theta_i(t)$ are assumed to be proportional to predicted changes in NPP and carbon accumulation. If NPP increases by 10 percent, for example, we assume that the merchantable part of trees in this system will grow 10 percent faster. This assumption implies that the merchantable part of a tree remains in the same proportion to its limbs and roots and to the carbon on the forest floor. The effect of productivity changes on annual growth rates is:

$$V_i(t, \theta_i(t)) = \int dV/ds\,(s, \theta_i(s))ds \qquad (7.1)$$

The stock of forests V_i at time t depends upon the cumulative effect of $\theta i(s)$, the changes of productivity on the annual growth of trees. During early periods, when climate is just beginning to have effects, these changes will have only a small impact on timber approaching maturity. After many years, however, $\theta_i(t)$ will have a larger impact as the cumulative effect of growing faster (slower) increases (decreases) the size of the stock.[1]

A simple equilibrium model is used to estimate a baseline case for timber markets. Although historical harvests in California have been strongly influenced by harvesting old growth, current harvests are largely restricted to second-growth forests. We begin by making some assumptions about the current inventory of forests in the state in each county. Ideally, we would prefer to begin with a detailed inventory that describes precisely how much acreage is in each age class. Unfortunately, such a detailed inventory is not available. The most we were able to secure is the aggregate acreage of timberland by county and the aggregate harvests in each county.

We assume that the private softwood forests in each county are currently fully regulated; that is, there are an equal number of hectares of softwood in every age category on private land. This assumption is likely to be accurate for the private lands across the state as a whole. However, it is unlikely that every county is fully regulated. Some counties probably have more land in younger age classes and some counties probably have more land in mature age classes. If this variation is random, it will not introduce a bias in the analysis.

We expect public lands throughout the state to have a disproportionate acreage of older trees, many of them in advanced maturity. The available data from the state suggest that almost two-thirds of the softwood forests (8.1 million of 12.9 million acres) in California are in public hands, and most of these lands are federal. Harvests from public lands, however, account for only 13 percent of state volume even though the public controls two-thirds of the forestland. We expect that this low harvest production level partially reflects low productivity lands in the high Sierras and the southern parts of the state. More importantly, however, the low harvest rates reflect the multiple purposes of public lands. Public forests are expected to encourage conservation and produce many other non-market services. Timber production is not the highest priority service on public land.

Because public harvests are not near their biological potential, we assume that changes in the biological potential of public lands will not necessarily lead to changes in harvests. In other words, the harvests on public lands will be determined by social choices that are largely independent of climate change. These choices do not reflect biological constraints such as productivity or biome changes on public lands but instead reflect preferences for non-market services over timber. This analysis consequently concentrates on the private management of softwoods in California.

We begin the analysis by calculating timber harvests in the baseline case. The baseline case projects what California harvests and net revenues will look like into the future if there is no climate change. When looking at climate change, we use the baseline case as a reference, and assume that the aggregate acreage of softwood forestland in each county remains constant in the baseline. Although there will be a substantial amount of development throughout the state, most new development is expected to take place in hardwood forests, shrub land, and grasslands. Private softwood forests are not expected to be affected by economic growth. In the baseline scenario, we also assume that forest productivity will remain constant; that is, we expect the trees to continue to grow at their current rates in each region of the state.

From available state data, we calculate the average productivity of forests in each county by dividing harvests by acreage. This analysis reveals two levels of average productivity across the state. The counties near the central valley, along the central coast, and in the northern interior tend to have slightly lower productivity. For this region, we estimate the following yield equation:

$$V(t) = \exp(12.31 - 145/t), \tag{7.2}$$

where V is the yield and t is the rotation length. The forests along the northern coast and the Sierras around Sacramento tend to have higher average growth rates. We fit the following parameters for this more productive region:

$$V(t) = \exp(12.54 - 145/t). \tag{7.3}$$

Both of these yield functions reflect the slow-growing species in California. The average age that maximizes sustained yield, for example, is 145 years in both equations. Of course, the functions above mask the yield functions of individual species such as the difference between Douglas fir, redwoods, and ponderosa pine. Unfortunately, the ecological model cannot discern these specific species within their biomes.

Given the yield functions, we next calculate the rotation age. Relying on the well-known Faustmann equation, we choose a rotation length that maximizes the present value of net revenues:

$$\text{Max}\,W(t) = [P^* V(t)\exp(-rt) - C]/[1 - \exp(-rt). \qquad (7.4)$$

where W is the present value of future rotations, P is the price of timber, V is the volume of timber at harvest time t, r is the interest rate, and C is the cost of planting trees. We assume that C is \$400 per acre, r is 4 percent real, and P is \$0.41 per board foot. Maximizing W with respect to t reveals that the optimal rotation length for California forests is 62 years. We assume that all private softwood lands are managed at this rotation length, meaning that all softwood stands on private lands are cut when they reach 62 years of age. Of course, not every hectare of forest will be harvested at this rotation length. Some stands on more fertile lands may be harvested at a younger age and stands on more marginal lands may be harvested later. The model is not able to distinguish land productivity beyond the regional data available.

The price per board foot varies across the state's counties from about \$0.35 to \$0.70 per board foot. The average for the state is \$0.41 per board foot. This price variation may simply be a measurement error. However, some of this variation may reflect permanent differences between counties due to access costs or other features of each county. We assume that the relative prices for timber across counties do not change over time or with climate change. We do explore a sensitivity analysis where we allow global prices to change with climate change.

To calculate the impact of climate change, we introduce the changes in growth rates and suitable softwood land that the ecological model predicts for each climate scenario. The changes are introduced as shown in equation (7.1). This leads to new growth rates in each decade from 2000 to 2100. The ecological model also predicts changes in the area of potential softwoods in each county. The model adjusts the acreage of softwoods that can be planted in proportion to predicted changes in aggregate acreage. The model then calculates the harvest and planting rates for each decade for each county and compares these rates to the baseline scenario. The model also calculates net revenue. Net revenue for each decade is equal to the harvest rates times the price per unit timber minus planting costs. The model calculates the change in net revenues in each decade from 2000 to 2100 and then calculates the change in net present value for the entire period.

Because the analysis is limited to the period between 2000 and 2100, the study ends abruptly in 2100. Impacts beyond 2100 are not calculated. Halting the analysis in 2100 ignores damages beyond 2100 that seem likely given the trends in the ecological data. Because limiting effects to the twenty-first century underplays some of the long-term harmful consequences of climate change on California's timber, future analysts may want to consider these longer-term effects and extend the analysis beyond 2100.

To examine the impact of changes in timber prices on California, we use the results of an earlier analysis of global timber impacts (Sohngen et al., 2002). That study begins with a set of climate changes that were evaluated by a global ecological model (Haxeltine and Prentice, 1996). The global ecological changes were then evaluated with a general equilibrium timber model that calculated timber prices over time. In this analysis, we begin with the predicted paths of timber prices from the global study. We explore how these price changes would affect California. The welfare impacts on timber suppliers in California are calculated by maximizing the Faustmann model with the new timber prices and estimating the new revenues. The welfare impacts on California consumers are calculated by estimating the change in consumer surplus under the demand for timber in the United States and then assuming that California's share is proportional to the state's share of the national population (Sohngen and Mendelsohn, 1998). The limitations of the price analysis are discussed below.

7.3 RESULTS

7.3.1 Within-State Analysis

We begin the analysis of within-state impacts by calculating the baseline: a projection of California timber harvests, planting, and net revenues from 2000 through 2100 without climate change. Given the assumption that the private forests are currently fully regulated and that they rely on Faustmann rotations, the model predicts that current harvest rates for timber can continue indefinitely into the future. California will generate approximately $920 million of timber sales per year from 2.3 billion board feet. We assume that private softwood will generate 1.93 billion board feet annually with gross revenues of $793 million. In the baseline, 4.8 million acres remain in private softwood timber and 81 500 acres of this land are harvested every year when the trees reach age 62. The rest of the state's timber supply (370 million board feet) will come from softwood on public lands and from hardwoods. Finally, we assume that real prices will not change in this first analysis and will remain at an average of $0.42 per board foot.

The baseline scenario predicts that gross timber revenues from private softwood will continue indefinitely each decade, yielding $793 million annually. Subtracting planting costs, the private forests in California will yield a net annual revenue of $760 million. Taking the present value of this stream of net revenue from 2000 through 2100 yields $18.8 billion. This is the present value of the net revenue in the twenty-first century for the 4.8 million acres of private softwoods under the baseline assumptions.

Now, we explore what effects climate change will have on timber net revenue in California. We make the assumption that climate change will affect only private softwood harvests. Public softwood harvests and private hardwood harvests are currently only a small fraction of potential harvests. At this time, these harvests are not dependent on ecological constraints – they are determined by other factors. Although this does not necessarily mean that public softwood and all hardwood harvests will not be affected by climate change, it does suggest that it is hard to know how changes in climate would alter these harvests. For example, the public could react to reductions in private timber supply by increasing harvests allowed on public lands or they could reduce public harvests in parallel with private reductions.

It is possible that omitting public softwoods and all hardwoods from the analysis will introduce a small bias. In general, climate is likely to have the same impact on public softwoods as it does on private softwood forests. The softwood forests lost from warming will largely turn into hardwoods. Thus, if harvests in these two forest types parallel what is happening to the forests, public softwood harvests will rise slightly, then decline sharply, and hardwood harvests will largely rise. Given that these two sources are a small fraction of current timber supply in California and that they move in opposite directions, the omission of these two sources does not seriously bias the analysis.

In this analysis, we explore how climate change would affect private softwoods. For each climate scenario, the ecological model predicts the amount of land in softwoods and the productivity for each county for each decade between 2000 and 2100. Table 7.1 presents the ecological results for three of these decades: 2020, 2060, and 2100. The two measures of productivity, NPP and net additions of carbon, reflect many of the same changes in the ecosystem except that the carbon measure includes fire. As a result, the two measures generally agree with each other but they have slightly different reactions to moisture. The carbon measure is more sensitive to moisture. For example, the carbon measure responds more positively to the wet Hadley scenario with predicted reductions in fire and more negatively to the dry PCM scenario with possible increases in fire. The two measures do not have a strong response to higher temperatures; they yield very similar outcomes

*Table 7.1   Intertemporal impact of climate change on ecological
parameters in California*

Scenario	Productivity (NPP)	Productivity (carbon)	Acreage (softwoods)
HadCM2			
2020	1.22	1.20	4698
2060	1.21	1.27	4357
2100	1.36	1.50	3922
PCM			
2020	1.05	1.17	4698
2060	1.20	1.04	4259
2100	1.22	0.96	3625
3.0°C, 0% P			
2020	1.00	0.99	4695
2060	1.07	1.02	4219
2100	1.03	1.06	3788
3.0°C, 18% P			
2020	1.02	1.01	4710
2060	1.14	1.08	4275
2100	1.16	1.18	3943
5.0°C, 0% P			
2020	1.01	0.99	4719
2060	1.08	1.02	4073
2100	1.01	1.03	3337
5.0°C, 30% P			
2020	1.05	1.02	4709
2060	1.20	1.12	4006
2100	1.22	1.23	3384

for the 3°C and 5°C scenarios with no precipitation change. In general, however, it appears that the climate scenarios increase productivity with the only major exception being the long-term carbon measure in the dry PCM scenario.

Table 7.1 also shows predictions of the acreage of softwood expected in each climate scenario. Climate change is expected to shrink the acreage of softwoods in California, turning a substantial fraction of softwood forests into hardwoods. The reduction increases with time in every scenario. Curiously, however, the reduction is only slightly sensitive to higher temperatures or less precipitation. It is this loss in softwood acreage that eventually reduces timber supply across the state.

The increase in productivity per hectare from climate change is not the same across the entire state. Table 7.2 presents the results in 2100 for each major region of the state's forests. (There is no effect in the southern portion of California because it is already too warm and dry to support softwoods even before climate change.) According to the NPP measure, uniform dry scenarios will result in the central coast losing productivity and the central valley, north coast, and Sacramento receiving only small benefits. The northern interior is the only region to benefit in the dry scenarios.

However, in the wetter uniform scenarios, only the central coast experiences a small effect and all other regions benefit. The PCM scenario resembles the uniform dry scenario in that the central coast is slightly damaged, but all other regions benefit. The Hadley scenario resembles the uniform wet scenarios except that even the central coast does well. The carbon measure treats the northern interior and Sacramento similarly to the NPP measure except for the PCM scenario, where it predicts large damages in both regions. The carbon measure results are much more optimistic about the central coast and the north coast than the NPP measure and slightly more pessimistic about the central valley. The carbon model's more optimistic predictions for the coastal regions and more pessimistic predictions for the central valley most likely reflect fire predictions for the wetter coast and drier valley.

As with the changes in productivity, the changes in softwood habitat vary across the state, as seen in Table 7.2. There are only small changes in softwood acreage in the central coast, but that region has very few acres of softwoods to begin with (98 000). In many of the uniform climate change scenarios, the remaining regions seem to have the same absolute losses except for slightly larger effects in the northern interior. However, in percentage terms, these effects fall much more heavily on the private softwoods in the central valley and Sacramento regions, with initial acreages of only 401 000 and 584 000 acres, respectively. In contrast, the percentage losses in the north coast and northern interior are lower because they contain so much more private softwood (1 960 000 and 1 850 000 acres, respectively). The Hadley scenario predicts much less harmful effects in the northern interior relative to the wet uniform change scenario because it predicts enormous increases in precipitation in the extreme northeastern corner of the state. These precipitation increases result in a substantial increase in softwoods in that location that offsets losses in the rest of the region. The PCM model predicts much more severe impacts on the northern coast and northern interior than the dry uniform scenario because it predicts substantial drying in the northern part of the state.

Given these changes in productivity and acreage, we next calculate what happens to the aggregate amount of land and average productivity in each

Table 7.2 Regional impact of climate change on ecological parameters in 2100

Scenario	Productivity (% Δ NPP)	Productivity (% Δ carbon)	Softwoods (Δ 000 acres)
HadCM2			
Central coast	34	114	3
Central valley	45	48	−292
North coast	12	17	−396
North interior	42	54	5
Sacramento	30	39	−291
PCM			
Central coast	−1	87	0
Central valley	23	−28	−201
North coast	12	38	−502
North interior	31	−17	−350
Sacramento	24	−27	−215
3.0°C, 0% P			
Central coast	−13	12	−1
Central valley	−1	−4	−274
North coast	4	18	−221
North interior	11	10	−348
Sacramento	2	3	−260
3.0°C, 18% P			
Central coast	7	32	0
Central valley	14	11	−277
North coast	19	22	−219
North interior	23	21	−209
Sacramento	16	17	−246
5.0°C, 0% P			
Central coast	−26	−7	−5
Central valley	0	−9	−330
North coast	3	16	−202
North interior	10	10	−701
Sacramento	0	2	−318
5.0°C, 30% P			
Central coast	−1	18	−1
Central valley	23	15	−327
North coast	12	24	−361
North interior	31	28	−519
Sacramento	24	26	−300

Table 7.3   Annual timber net revenue (millions of dollars/year)

Scenario	2020	2060	2100
NPP model			
HadCM2	775	866	869
PCM	784	772	784
3°C, 0% P	775	744	680
3°C, 18% P	778	761	726
5°C, 0% P	781	757	653
5°C, 30% P	785	792	743
Carbon model			
HadCM2	780	800	836
PCM	791	891	846
3°C, 0% P	786	709	636
3°C, 18% P	788	709	667
5°C, 0% P	787	715	601
5°C, 30% P	792	739	674
Baseline	760	760	760

climate scenario in each decade. Initial impacts are attributable strictly to changes in productivity. Harvests increase slightly because the standing trees are growing more rapidly. The reduction in softwood acreage reduces new land planted immediately but it takes many decades to see this effect on harvests. Table 7.3 shows the effects on net revenue in 2020, 2060, and 2100.

Relative to the baseline of $760 million per year, the net revenues are higher in 2020 for every scenario. However, in all the uniform change scenarios, net revenues strictly decline with time. Effects diverge across climate scenarios by 2060. Using NPP as a measure of productivity, net revenues under the uniform wet scenarios are higher through 2060 and then fall off and become damages. The net revenue in the uniform dry scenarios has already fallen below the baseline by 2060, and continues to shrink with time. Interestingly, the PCM scenario is beneficial even though it is dry, and the Hadley scenario is the most beneficial of all the climate change scenarios. By 2100, the only two scenarios that are beneficial are the two GCMs. Using the carbon measure of productivity, the two GCM scenarios remain the most beneficial in the far future, but in this case, the PCM scenario is actually more beneficial than the Hadley. The carbon measures also imply that all the uniform change scenarios lead to falling net revenue over time. With the carbon measure, however, these paths change only slightly with more severe temperature increases or with less precipitation.

Table 7.4   *Present value of impact of climate change on timber (millions of dollars) (baseline $18.8 billion)*

	HadCM2	PCM	3.0°C, 0% P	3.0°C, 18% P	5.0°C, 0% P	5.0°C, 30% P
				Climate scenario		
NPP model						
State	755.9	347.7	57.6	180.9	169.2	373.5
	(4.0%)	(1.9%)	(0.3%)	(1.0%)	(0.9%)	(2.0%)
Central coast	19.6	13.9	6.6	11.7	4.8	13.8
	(4.7%)	(3.3%)	(1.6%)	(2.8%)	(1.1%)	(3.3%)
Central valley	29.8	5.2	−32.0	−19.4	−30.9	−5.3
	(2.4%)	(0.4%)	(−2.6%)	(−1.6%)	(−2.5%)	(−0.4%)
North coast	259.1	106.5	48.6	85.5	97.4	153.6
	(2.7%)	(1.1%)	(0.5%)	(0.9%)	(−1.0%)	(1.6%)
North interior	402.5	161.5	37.1	87.2	88.9	177.1
	(7.4%)	(3.0%)	(0.7%)	(1.6%)	(1.6%)	(3.2%)
Sacramento	44.9	60.6	−2.7	15.9	9.1	34.3
	(2.2%)	(2.9%)	(0.1%)	(0.8%)	(0.4%)	(1.7%)
Carbon model						
State	599.6	1,188.7	160.5	213	195.6	361.6
	(3.2%)	(6.3%)	(0.8%)	(1.1%)	(1.0%)	(1.9%)
Central coast	75.8	152.9	37.3	43	35.3	42.1
	(18.2%)	(36.6%)	(8.9%)	(10.3%)	(8.5%)	(10.1%)
Central valley	50.3	−53.8	−20.2	−8.5	−21.6	−1.4
	(4.1%)	(−4.3%)	(−1.6%)	(−0.7%)	(−1.7%)	(−0.1%)
North coast	4.9	1.140.7	125.6	96.3	148.7	188
	(0.1%)	(11.9%)	(1.3%)	(1.0%)	(1.6%)	(2.0%)
North interior	419.2	−49.6	5.8	50.8	24.3	100.5
	(7.7%)	(−0.9%)	(0.1%)	(0.9%)	(0.4%)	(1.8%)
Sacramento	49.3	−1.6	12	31.4	8.8	32.5
	(2.4%)	(−0.1%)	(0.6%)	(1.5%)	(0.4%)	(1.6%)

In Table 7.4, we examine the present value of the streams of net revenue displayed in Table 7.3. Using a 4 percent real interest rate, we calculate the change in the present values of net revenue caused by climate change. These calculations take the difference in the stream of net revenue under each climate scenario relative to the baseline. The results, as shown in Table 7.4, reveal that the effect of climate change on California timber is positive. In every scenario, the present value of net revenues increases. The benefit from warming ranges from $58 million (0.3 percent) in the 3°C, 0 percent scenario with carbon to $1.2 billion (6.3 percent) in benefits in the PCM scenario using NPP. The present value calculation implies that the benefits of the early increases in productivity outweigh the damages from the eventual loss of softwood habitat. The two GCM scenarios are the most beneficial, with benefits ranging from $348 million to $1.2 billion. The uniform wet scenarios are better than the dry scenarios and the 5°C uniform scenarios are better than the 3°C scenarios. In general, with the dry scenarios, the NPP measure is more optimistic, and with the wet scenarios, the carbon measure is more optimistic. Again, this difference in predictions most likely results from the fact that the carbon measures include the effects of fire.

Although warming raises timber net revenue across the state, it does not do so in every region. The Central Valley is likely to be damaged by warming in every scenario except the Hadley. The PCM scenario with the carbon model also leads to damages in the northern interior and Sacramento regions, although these effects are small in percentage terms. The PCM scenario predicts curiously large benefits in the central (36 percent) and north (12 percent) coast in the carbon measure and the Hadley predicts large benefits in the northern interior (7–8 percent). In general, however, the benefits from warming tend to have a small effect in each region, in the range between 1 percent and 3 percent. Across the uniform climate change scenarios, the warmer wetter scenarios lead to more benefits (or less damages).

7.3.2   Analysis of Price Changes

The results in Tables 7.3 and 7.4 are based on changes in California forest productivity. In this first analysis, we have assumed that the prices of timber will remain constant. Of course, it is unlikely that timber supplies will change from 1 percent to 6 percent and not change prices. The problem is that we cannot assume that just because California supply changes, global supplies will change proportionately. The California supply of timber is a poor measure to use to predict global timber prices.

The second analysis examines the impact of global price changes caused by climate change. The climate scenario in the global model is not one of the scenarios used in this book. However, we select the UIUC (University of Illinois at Urbana-Champaign General Circulation Model) climate scenario from the global study, which is similar to the scenario from HadCM2. The purpose of this analysis is not to contrast what happens to California forests with what happens in forests in the rest of the world, in which case one should use the same climate scenario for both studies. The purpose is to examine the impact on California of plausible changes in global timber prices.

A different ecological model was used to generate the global forest effects (Haxeltine and Prentice, 1996). In a model comparison test, the models used in both the national and global studies were shown to be similar

(VEMAP Participants, 1995). In general, the ecological models predict that a warmer, wetter, CO_2-enriched world will cause global forests to be more productive (up to a point; actually they show NPP peaking around $+1°C$). The global model, like the California model, predicts increases in forest productivity over time. However, the global model predicts larger initial benefits as short-rotation semi-tropical forests respond quickly to carbon fertilization effects. The global model also predicts that the benefits will continue to increase over the century as forests increase in size globally. Much of this increase in forest area will occur in the polar regions. These long-term benefits seen elsewhere in the world do not occur in California. In California, softwoods are predicted to shrink because they are on the southern edge of their biome. Thus, for the world, ecological-economic models predict that timber supply will increase, although in California, timber supply will fall over time.

Given the increase in global timber productivity, the economic timber model predicts that timber prices will fall over time as a result of climate change (Sohngen et al., 2002). We change the California prices in each county proportionally with the changes in global prices.

Including the changing prices in the Faustmann formula reveals that foresters would not change optimal rotations in response to the price changes. Because there are few competing land uses in the regions where the forests are located, the acreage of the forests would not likely change either. Consequently, the price changes are not likely to cause a large change in the supply of timber. The welfare effect on forest producers is likely to be equal to the change in price times the quantity of production.

In Tables 7.5 and 7.6, the changes in global prices predicted by Sohngen et al. (2002) are used to predict what will happen to California. California prices fall over time because of the predicted increases in global production. Including these prices in the baseline scenario reveals that even if climate change had no effect on production in California, the resulting net revenues would still fall because of falling prices. Adding the changing prices to the changing production levels gives very different results in Table 7.5 from those presented in Table 7.3. When prices are held constant (Table 7.3), net revenues increase with the two GCM scenarios and increase and then fall with the uniform scenarios. With prices falling over time (Table 7.5), net revenues fall in every climate scenario over time. Relative to the baseline with no climate change, there are damages for producers in every decade that increase steadily with time.

In Table 7.6, we take the present value of the stream of annual damages reported in Table 7.5 associated with including global prices in the analysis. Including global price effects, climate change has a negative effect on California timber producers. Despite the beneficial effects on production in

Table 7.5   Annual timber net revenue with falling global prices (millions of dollars/year)

Scenario	2020	2060	2100
NPP model			
HadCM2	694	669	644
PCM	703	596	581
3°C, 0% P	695	573	503
3°C, 18% P	697	587	538
5°C, 0% P	699	584	484
5°C, 30% P	703	611	551
Carbon model			
HadCM2	699	617	620
PCM	709	688	628
3°C, 0% P	704	547	470
3°C, 18% P	706	546	494
5°C, 0% P	705	551	445
5°C, 30% P	709	570	499
Baseline	760	760	760
Baseline with falling prices	681	586	562

California, climate change causes price reductions that lead to damages in every scenario. The damages range from $0.1 billion to $1.4 billion. Every region has net damages except for the central coast in the carbon model. Because prices reduce net revenues equally in all regions, the proportional damages are similar across the remaining regions except for slightly larger damages in the central valley. Note that with falling prices, the absolute size of damages is much larger in the regions with more timber (north coast and northern interior).

This sensitivity analysis reveals that California timber owners are more vulnerable to global price reductions than to local reductions in productivity. If there are no changes in prices, climate effects are beneficial for timber producers and could even deliver benefits as high as $1 billion. However, if prices fall because of global increases in forest productivity, California timber producers are likely to suffer damages of more than $1 billion.

However, the very same price reductions that adversely affect timber suppliers provide large benefits to California consumers. The price elasticity of demand is steeper than the price elasticity of supply, so that price declines lead to more consumer surplus gained than producer surplus lost (Sohngen et al., 2002). Further, California consumes much more timber than it produces. The state exports 341 million board feet to other western states but imports 1278 million board feet for a net import of 936 million board feet.

*Table 7.6   Present value of climate change impact on timber with falling
global prices (millions of dollars)*

	Climate scenario					
	HadCM2	PCM	3.0°C, 0% P	3.0°C, 18% P	5.0°C, 0% P	5.0°C, 30% P
NPP model						
State	−984.8	−1311.1	−1547.1	−1448.2	−1454.0	−1291.0
	(5.2%)	(7.0%)	(8.2%)	(7.7%)	(7.7%)	(6.9%)
Central coast	−20.4	−24.1	−29.9	−25.9	−31.4	−24.1
	(4.9%)	(5.8%)	(7.2%)	(6.2%)	(7.5%)	(5.8%)
Central valley	−80.6	−100.3	−130.0	−119.8	−128.7	−108
	(6.5%)	(8.1%)	(10.5%)	(9.7%)	(10.4%)	(8.7%)
North coast	−594.4	−719.2	−769.9	−740.1	−728.8	−684
	(6.2%)	(7.5%)	(8.0%)	(7.7%)	(7.0%)	(7.1%)
North interior	−148.3	−340.0	−439.7	−399.8	−397.4	−147.3
	(2.7%)	(6.2%)	(8.0%)	(7.3%)	(7.3%)	(7.1%)
Sacramento	−141.1	−127.5	−177.5	−162.6	−167.8	−147.3
	(6.8%)	(6.2%)	(8.6%)	(7.9%)	(8.1%)	(7.1%)
Carbon model						
State	−1084.6	−616.1	−142.5	−1381.8	−1395.9	−1261.9
	(5.8%)	(3.3%)	(7.6%)	(7.4%)	(7.4%)	(6.7%)
Central coast	26.4	90.9	−3.5	1.1	−5.1	−0.3
	(6.3%)	(21.8%)	(0.8%)	(0.3%)	(1.2%)	(0.1%)
Central valley	−62.2	−149.1	−118.8	−109.3	−119.5	−103.3
	(5.0%)	(12.1%)	(9.6%)	(8.8%)	(9.7%)	(8.4%)
North coast	−785.2	134.5	−680.5	−703.1	−661.3	−628.8
	(8.2%)	(1.4%)	(7.1%)	(7.30%)	(6.9%)	(6.5%)
North interior	−127.1	−514.1	−459.1	−423	−444.3	−383.8
	(2.3%)	(9.4%)	(8.4%)	(7.7%)	(8.1%)	(7.0%)
Sacramento	−136.4	−178.4	−162.8	−147.3	−165.8	−146.4
	(6.6%)	(8.6%)	(7.9%)	(7.1%)	(8.0%)	(7.1%)

Total California production is therefore 1598 million board feet, but total consumption is 2535 million board feet.

The consumer surplus benefits from lower prices increase over time with population growth. We examine the slow growth and fast growth projections made in Chapter 2. As can be seen in Table 7.7, there are large benefits to consumers from the lower prices. In the slow growth scenario, the benefits are estimated to be $490 million in 2020, growing to $1 billion in 2060 and beyond. In the fast growth scenario, the 2020 estimates are similar but benefits grow to $1.2 billion by 2060 and $1.5 billion by 2100. The present value of this stream of benefits to consumers through 2100 is equal

Table 7.7   Impact of global warming on California consumers (millions of dollars)

	Present value	Annual impact		
		2020	2060	2100
Slow growth				
State	13 156	490	1000	1000
South	8 077	301	614	614
Central coast	2717	101	206	206
Central valley	1719	64	131	131
North coast	270	10	21	21
North interior	173	6	13	13
Sacramento	200	7	15	15
Fast growth				
State	14 280	486	1157	1537
South	8767	298	710	944
Central coast	2949	100	239	317
Central valley	1866	64	151	201
North coast	293	10	24	32
North interior	188	6	15	20
Sacramento	217	7	18	23

to $13 billion in the slow growth scenario and $14 billion in the fast growth scenario.

Consumers all across the state will gain from the lower prices, but counties will gain in proportion to population. Even counties that have no timber production will enjoy gains in consumer surplus. This is especially evident in the southern urban counties, which gain about 60 percent of the benefits, and in the San Francisco region, which enjoys another 20 percent. The remainder of the benefits is spread across the more rural parts of the state. Because the timber-growing northern and mountain regions have relatively low populations, they receive only a small fraction (5 percent) of the statewide consumer benefits.

7.4   CONCLUSIONS

This chapter presents estimates of the welfare impacts of climate change on California timber. We explored two possible consequences of climate change: it can change local productivity and it can change global prices. Several climate scenarios were used to explore local productivity effects to understand what might happen with a range of temperature and precipitation changes. A dynamic ecological model then predicted how acreage, fire, and productivity would shift across biomes over time. An economic model then predicted how these ecological changes would change county level timber supply each decade. Assuming no change in prices, the economic model then predicted net revenues.

The ecological model predicts that timber productivity will increase but that acreage of productive softwoods will decline. The economic model predicts that statewide net revenues will increase at first in all scenarios because of the increase in productivity. As time passes, some of the scenarios predict that net revenue will subsequently decline, even below baseline levels, whereas others predict continued increases in net revenues. The present value increases as a result of all these changes. The early increases in net revenue outweigh the later declines in all scenarios when discounted. Changes in local productivity from climate change will provide benefits to California timber owners at first but then losses in the latter half of the century. The hotter and wetter the scenario, the bigger the net increase in benefits to forest owners. The impacts, however, are not uniform across the state. Net forest income will increase more in the alpine and northern areas of the state.

The analysis also examines the impact of changing global timber prices. Because global timber productivity is projected to increase, global timber prices are projected to fall over time. These falling prices will cause net revenues for forest owners in California to fall dramatically over time in every scenario. The positive impacts of local productivity changes are overwhelmed by the falling prices, leading to net present value losses to timber producers of more than $1 billion in many cases. The reductions in global prices, however, lead to large gains for consumers. The predicted increase in the net present value of welfare to California consumers is in the neighborhood of $14 billion. Taking all these changes into account, warming is likely to provide large net benefits to California by providing lower prices for timber consumers.

This research indicates a wide range of possible effects on California's timber resource. Part of the uncertainty lies in the climate scenarios themselves because the state may experience mild or severe warming and dry or wet conditions. This range of climate outcomes will lead to different predictions of local productivity changes and of biome changes. Most of the scenarios imply that productivity changes are likely to be positive and the biome changes will be negative for timber. The research, however, highlights that timber resources in California are even more sensitive to changes in global timber prices. If forests in the rest of the world increase in productivity as much as many models predict, prices for timber will fall, leading to losses for California producers and gains for consumers. Depending on these price changes, the benefits to consumers could well dwarf the losses to the state's producers. The effects of climate change outside the state may be far more important than the observed effects in the state.

Several limitations of the current study could be addressed in future work. As part of the unifying conditions set forth in this study to make the results consistent across sectors, the study was limited to the period between 2000 and 2100. It is clear that warming will continue to erode local productivity into the future, so this effect will continue to get worse after 2100. In addition, the study explores the importance of global prices but relies on a global model for price predictions that are not perfectly consistent with the California model. The global model relied on a different climate scenario and a different ecological model. Given its clear importance, more research into possible global price paths is worthwhile.

NOTE

1. The studies in the literature actually make a range of assumptions about how climate affects productivity so that the specific form of equation (7.1) depends on the underlying assumptions.

REFERENCES

Binkley, C.S. 1988. A case study of the effects of CO_2-induced climatic warming on forest growth and the forest sector: B. Economic effects on the world's forest sector. In *The Impact of Climatic Variations on Agriculture*, M.L. Parry, T.R. Carter, and N.T. Konijn (eds). Kluwer, Dordrecht, pp. 197–218.

Bowes, M. and R. Sedjo. 1993. Impacts and responses to climate change in forests of the MINK region. *Climatic Change* **24**: 63–82.

Burton, D.M., B.A. McCarl, N.M. de Sousa, D.M. Adams, R.A. Alig, and S.M. Winnet. 1998. Economic dimensions of climate change impacts on southern forests. In *Ecological Studies 128: The Productivity and Sustainability of Southern Forest Ecosystems in a Changing Environment*, S. Fox and R. Mickler (eds). Springer-Verlag, New York, pp. 777–94.

California Department of Finance. 2003. *California Statistical Abstract*. Available at http://www.dof.ca.gov/HTML/FS_Data/STAT-ABS/Sa_home.htm.

Emanuel, W.R., H.H. Shugart, and M.P. Stevenson. 1985. Climate change and the broad-scale distribution of terrestrial ecosystem complexes. *Climatic Change* **7**: 29–43.

Gitay, H., S. Brown, W. Easternling, and B. Jallow. 2001. Ecosystems and their goods and services. In *Climate Change 2001: Impacts, Adaptation, and Vulnerability. Contribution of Working Group II to the Third Assessment Report of the Intergovernmental Panel on Climate Change*, J.J. McCarthy, O.F. Canziani,

N.A. Leary, D.J. Dokken, and K.S. White (eds). Cambridge University Press, Cambridge, UK, pp. 235–42.

Haxeltine, A. and I.C. Prentice. 1996. BIOME3: An equilibrium terrestrial biosphere model based on ecophysiological constraints, resource availability, and competition among plant functional types. *Global Biogeochemical Cycles* 10(4): 693–709.

Howard, J. 2001. US timber production, trade, consumption, and price statistics 1965–1999. USDA, Forest Product Laboratory, FPL-RP-595.

Joyce, L.A., J.R. Mills, L.S. Heath, A.D. McGuire, R.W. Haynes, and R.A. Birdsey. 1995. Forest sector impacts from changes in forest productivity under climate change. *Journal of Biogeography* 22: 703–13.

Loomis, J. and J. Crespi. 1999. Estimated effects of climate change on selected outdoor recreation activities in the United States. In *The Impact of Climate Change on The United States Economy*, R. Mendelsohn, and J. Neumann (eds). Cambridge University Press, Cambridge, UK, pp. 289–314.

McCarl, B.A., D.M. Adams, R.D. Alig, D. Burton, and C. Chen. 2000. The effects of global climate change on the US forest sector: Response functions derived from a dynamic resource and market simulator. *Climate Research* 15(3): 195–205.

Melillo, J.M., A.D. McGuire, D.W. Kicklighter, B. Moore III, C.J. Vorosmarty, and A.L. Schloss. 1993. Global climate change and terrestrial net primary production. *Nature* 363: 234–40.

Mendelsohn, R. and M. Markowski. 1999. The impact of climate change on outdoor recreation. In *The Impact of Climate Change on The United States Economy*, R. Mendelsohn and J. Neumann (eds). Cambridge University Press, Cambridge, UK, pp. 267–88.

Neilson, R.P. and D. Marks. 1994. A global perspective of regional vegetation and hydrologic sensitivities from climate change. *Journal of Vegetation Science* 5: 715–30.

Pendleton, L. and R. Mendelsohn. 2000. Estimating recreation preferences using hedonic travel cost and random utility models. *Environmental and Resource Economics* 17: 89–108.

Shugart, H.H., M.Y. Antonovsky, P.G. Jarvis, and A.P. Sandford. 1986. CO_2, climatic change, and forest ecosystems. In *The Greenhouse Effect, Climatic Change and Ecosystems*, B. Bolin, B.R. Doos, J. Jager, and R.A. Warrick (eds). Wiley, Chichester, UK, pp. 475–521.

Sohngen, B. and R. Mendelsohn. 1998. Valuing the market impact of large scale ecological change in a market: The effect of climate change on US timber. *American Economic Review* 88: 686–710.

Sohngen, B., R. Mendelsohn, and R. Sedjo. 1999. Forest conservation, management, and global timber markets. *American Journal of Agricultural Economics* 81(1): 1–13.

Sohngen, B., R. Mendelsohn, and R. Sedjo. 2002. A global model of climate change impacts on timber markets. *Journal of Agricultural and Resource Economics* 26: 326–343.

Solomon, A.M. 1986. Transient response of forest to CO_2-induced climate change: Simulation modeling experiments in eastern North America. *Oecologia* 68: 567–79.

VEMAP Participants. 1995. Vegetation/Ecosystem Modeling and Analysis Project (VEMAP): Comparing biogeography and biogeochemistry models in a continental-scale study of terrestrial ecosystem responses to climate change and CO_2 doubling. *Global Biogeochemical Cycles* 9(4): 407–37.

8.   Changes in runoff

Norman L. Miller, Kathy E. Bashford, and Eric Strem

8.1   INTRODUCTION

A number of investigations of California hydrologic response have focused on changes in streamflow volumes resulting from climate change (for example, Revelle and Waggoner, 1983; Gleick, 1987; Lettenmaier and Gan, 1990; Jeton et al., 1996; Miller et al., 1999; Wilby and Dettinger, 2000; Knowles and Cayan, 2001). Using historical data, Revelle and Waggoner (1983) developed regression models to estimate the sensitivity of streamflow in major basins to climate change. Gleick (1987) used a modified upper and lower basin water budget model (Thornthwaite and Mather, 1955) for the Sacramento drainage directly forced by precipitation and temperature output from three general circulation models (GCMs). Lettenmaier and Gan (1990) used precipitation and temperature from three GCM scenarios to force process-based, basin-scale, water budget models (Anderson, 1973; Burnash et al., 1973) with three to five elevation band defined sub-basins at four basins (North Fork American, Merced, McCloud, and Thomes Creek) in the Sacramento–San Joaquin drainage. Jeton et al. (1996) ran a distributed parameter precipitation runoff model (Leavesley et al., 1983) to evaluate the North Fork American and East Fork Carson rivers using specified incremental temperature and precipitation as uniform climate change scenarios. Miller et al. (1999) dynamically downscaled a GCM projection via a regional climate model and used the output as forcing to process-based hydrologic models (Beven and Kirkby, 1979; Leavesley et al., 1983) in the North Fork American River and the north coastal Russian River. Knowles and Cayan (2001) used historical precipitation and a single GCM projection of temperature that was statistically interpolated to a 4 km resolution as input forcing to a modified version of the Burnash et al. (1973) soil moisture accounting model (Knowles, 2000) for the entire Sacramento–San Joaquin drainage.

In general, each of these studies suggested that Sierra Nevada snowmelt-driven streamflows are likely to peak earlier in the season under global warming. A key finding of these studies was that the greatest influences on streamflow sensitivity to climate change are the basin elevation and the location of the freezing line. To further understand the likelihood of potential shifts in the timing and magnitude of California streamflow and related hydrologic response, we analyzed six major watersheds forced by the climate change scenarios described in Chapter 4. The results of this study were then used in the assessment of how California's water management system would respond to climate change (including changes in demand for water; Chapter 10).

8.2 METHODS

This study focused on determining the range of the effects of projected climate change scenarios for assessing California hydrologic response. Streamflow sensitivities for the watersheds studied were related to a larger set of watersheds representing the entire Sacramento–San Joaquin drainage and were applied to water demand and allocation simulations.

Streamflow simulations in this study were based on the application of the National Weather Service River Forecast System. This system has been calibrated to California using the Sacramento Soil Moisture Accounting (SAC-SMA) Model (Burnash et al., 1973) coupled to the snow accumulation and ablation Anderson Snow Model (Anderson, 1973). The SAC-SMA has two upper zone storage compartments (free and tension) and three lower zone storage compartments (free primary, free secondary, and tension). Tension zone storage is depleted only by evapotranspiration processes; free zone water drains out as interflow and baseflow. The SAC-SMA was chosen primarily because it depends on only two variables, precipitation and temperature, and because it is the operational model of the NWS. It has been used in previous climate change sensitivity studies (Lettenmaier and Gan, 1990; http://www.energy.ca.gov/pier/final_project_reports/500-03-058cf.html) with an assumption of geomorphologic stream channel stationarity. Assuming fixed channel geometry requires that climate change simulations be based on perturbations about the historical data period for which the calibration was performed and verified (Lettenmaier and Gan, 1990). Historical temperature and precipitation time series for 30 years (1963–92) are sufficiently long for a representative climatology and are available at 6 h timesteps for each basin. The snow-producing basins were delineated into upper and lower basins with separate input forcing to account for the elevation, land surface characteristics, and climate differences.

The six representative headwater basins with natural flow selected for analysis (Figure 8.1) are the Smith River at Jed Smith State Park, Sacramento River at Delta, Feather River at Oroville Dam, American River at North Fork

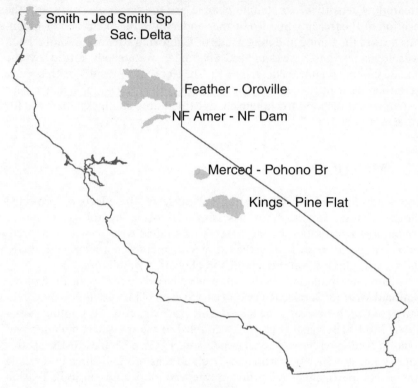

Figure 8.1   Location of the six study basins

Dam, Merced River at Pohono Bridge, and Kings River at Pine Flat Dam. Table 8.1 shows the basin size, location, percent area, and centroid of each upper and lower sub-basin. The Smith is a very wet coastal basin that does not significantly accumulate seasonal snowpack. The Sacramento is a mountainous northern California basin with a small amount of seasonal snow accumulation. The Sacramento provides streamflow for the north and northwest drainage regions into the Central Valley. The Feather and the Kings represent the northernmost and southernmost Sierra Nevada basins, and the Kings and Merced are the highest elevation basins. The American is a fairly low-elevation Sierra Nevada basin, but has frequently exceeded flood stage, resulting in substantial financial losses. This set of study basins provided sufficient information for a spatial estimate of the overall response of California's water supply (excluding the Colorado River) and helped indicate the potential range of impacts. This study modeled natural (or 'unmanaged') flows feeding into the state's water management system.

Table 8.1 Basin area, stream gauge coordinates, percent sub-basin area, and elevation

	Smith	Sacramento	Feather	American	Merced	Kings
Area, km^2	1706	1181	9989	950	891	4292
Gauge latitude (W)	41° 47′ 30″	40° 45′ 23″	39° 32′ 00″	38° 56′ 10″	37° 49′ 55″	36° 49′ 55″
Gauge longitude (N)	124° 04′ 30″	122° 24′ 58″	121° 31′ 00″	121° 01′ 22″	119° 19′ 25″	119° 19° 25″
Percent upper	0	27	58	37	89	72
Upper centroid, m		1798	1768	1896	2591	2743
Lower centroid, m	722	1036	1280	960	1676	1067

Historical precipitation and temperature input forcing to the hydrologic models is based on the archived NWS 6 h mean-area precipitation (MAP) and mean-area temperature (MAT) for each upper and lower basin. The NWS also provided historical daily streamflow data for the stream gauges at the outlet of each of the six basins.

NOAA's California Nevada River Forecast Center calibrated and verified each basin using parts of the 6 h and daily 1950 to 1993 precipitation, temperature, and streamflow time series. A 30-year climatological verification for the 1963–92 period using these calibration parameters was performed because it is the most complete and is close to the National Climatic Data Center 1961–90 climatology. Comparison of the observed to simulated climatological streamflow for the 1963–92 period resulted in monthly streamflow correlation coefficients greater than 0.95 for each of the six basins.

8.2.1   Incremental Perturbations

Streamflow was estimated by imposing the six incremental sets of constant temperature shifts ($T_{shift,incr}$) and precipitation ratios ($P_{ratio,incr}$) discussed in Chapter 4 onto the historical MAT and MAP time series. Adjusted 6 h temperature and precipitation input data were calculated by uniformly adding the temperature shift, and multiplying by the precipitation ratio, for each temperature and precipitation time series ($T_{incr}(t) = T(t)_{hist} + T_{shift,incr}$; $P_{incr}(t) = P(t)_{hist} * P_{ratio,incr}$). For each of the six incremental changes, daily streamflow ($Q_{day,incr}$) was simulated at each of the representative basins. From these daily streamflow outputs, monthly mean-daily streamflow, in cubic meters per second daily, was computed for October 1963 to September 1992. Monthly climatological means were computed as the monthly mean-daily streamflow for each calendar month over the 30-year period ($Q_{month,incr}$). Monthly means were also calculated for each observed 1963–92 streamflow time series to provide historical mean-monthly baseline climatologies ($Q_{month,hist}$).

8.2.2   Scenario Perturbations

The scenario perturbation studies used the two GCMs and downscaling technique described in Chapter 4. From these coupled atmosphere–ocean GCM simulations, two 30-year periods (2010–39 and 2050–79) and one 20-year period (2080–99) were used. Monthly temperature shifts and precipitation ratios derived from the mean-area basin climatologies were imposed on the historical 1963–92 temperature and precipitation time series as in the incremental studies.

GCM-based monthly mean-area precipitation and temperature were determined for each upper and lower sub-basin as the mean of the 10 km gridded temperature and precipitation within the sub-basin. Similarly, a set of basin mean-area historical monthly MAP and MAT time series were derived from the available 10 km historical data for 1963–92. Baseline climatological monthly MAP and MAT values were calculated from these 30-year records.

A ratio (shift) between the monthly basin mean area MAP_{scen} (MAT_{scen}) climatologies for the projected time periods and the monthly baseline historical precipitation (temperature) climatologies was computed. These climate scenario precipitation ratios ($P_{ratio,scen}$) and temperature shifts ($T_{shift,scen}$) were used to adjust the archived NWS observed time series in a similar manner as the constant incremental values, but in this case monthly adjustments were made. The imposed climate scenario mean-area precipitation and temperature time series were used as input to the hydrologic models as described.

8.2.3 Limitations

Although superimposing the temperature shifts and precipitation ratios from GCM predictions onto a historical time series is a valid approach, it does not capture the possibility that climate change may alter temperature and precipitation variance. Nor does it account for changes in the probability of extreme events. The potential error associated with these assumptions increases over time the further one gets from current conditions. So the forecasts are probably reliable for the next few decades but grow more uncertain toward the end of the century.

The assumption of fixed land use results in surface characteristics in both the GCMs, and the SAC-SMA also may not adequately represent future energy and water budgets. Using the SAC-SMA with a fixed evapotranspiration (ET) demand curve cannot explicitly yield ET climate change response with temperature, which is important during the dry March through August period. This implies that the simulated streamflow is higher than it should be during these periods of ET depletion. This effect is not significant during the snow accumulation period and is of less magnitude than GCM uncertainties such as cloud fraction.

8.3   RESULTS

Analysis of temperature, precipitation, snow-to-rain shifts with elevation, snowmelt, and streamflow was based on the mean monthly climatologies.

Shifts in the cumulative streamflow and exceedence probabilities of peak streamflow were based on the perturbation of the daily 30-year time series and annual peakflow.

8.3.1 Temperature

Because we are especially interested in the effects of climate change on water flows, we examine the annual temperature cycle at three of the head-water study basins (Sacramento, American, and Merced). We compare the two GCM projections (Hadley and PCM) superimposed on the NWS observed data. The simulated temperature climatologies generally follow the historical seasonal trends, with quasilinear increases with time. The greatest increases from the baseline are during June to August (JJA) season and in January, with the largest increase during Hadley 2080–99, followed by Hadley 2050–79, then PCM 2080–99. The monthly temperature shift ranges are 0.53°C to 4.70°C for the Hadley and −0.14°C to 3.00°C for the PCM.

The sensitivity of snowmelt to temperature increases depends on how many degrees the baseline temperature is below freezing during these months. The high-elevation upper Merced and Kings basins, where the December to February temperatures are several degrees below freezing, are less sensitive to small temperature increases than the upper American basin, where the winter temperatures are about 1°C below freezing. The increased summertime heating will increase ET, reducing soil moisture storage and streamflow.

8.3.2 Precipitation

We make a similar comparison for the mean-monthly precipitation for the same three headwater basins (Sacramento, American, and Merced). The simulated future climate mean-monthly precipitation volumes do not follow the historical cycle closely. The warm, wet Hadley increases in monthly amounts from November to March, and generally shifts the maximum precipitation by about one month later in the year. The PCM total annual precipitation is close to the historical precipitation; however, precipitation decreases from November to December and again during March and April for the 2050–79 and 2080–99 mean climates. In January, the 2050–79 period shows a large increase, whereas the other months show a significant decrease.

The wet Hadley projection consistently shows higher ratios than does the drier PCM projection. The Hadley has a minimum wet season precipitation ratio of 0.89 in December of 2010 to 2039 and a maximum of

2.04 times the baseline during February of 2080 to 2099. The PCM precipitation ratios have a much smaller range, with a wet season minimum of 0.48 times the baseline in November of 2080 to 2099 and a maximum of 1.16 times the baseline in January of 2050 to 2079. The range of PCM precipitation ratios is less than the high incremental precipitation ratio (1.3) and shows a decrease in precipitation. The Hadley exceeds the high incremental ratio in the Merced and Kings basins for 2050–79, and in all basins for 2080–99.

8.3.2.1 Snow-to-rain ratios

The snow-to-rain ratios vary significantly with latitude, and most importantly with the level of the lower and upper basins. In this study, the elevation band partition was based on the historical snow accumulation line. The Anderson Snow Model's area elevation curve and the snow-to-rain line determine the percentage of the sub-basin's area that is covered by snow and how that snow-covered area changes over time. This removes the need for a large number of elevation band sub-basins for determining the percent snow and percent rain within each sub-basin area.

The lower sub-basins typically have minimal to no accumulation and the upper sub-basins have the majority of the accumulated snow. High-elevation sub-basins (for example Upper Merced at 2591 m) see higher snow accumulations and later-season runoff than do the lower elevation sub-basins (for example Upper Sacramento at 1798 m) for the climate change scenarios. The elevation-dependent snow-to-rain ratios shift for each projection (Figure 8.2). Although the Hadley projections show a significant increase in total precipitation, and the PCM projections show reduced precipitation, both cases show a significant reduction of the snow-to-rain ratio.

8.3.2.2 Snow water equivalent

Figure 8.3 shows the change relative to the baseline snow water equivalent (SWE) of the snowpack for the snow-producing upper and lower sub-basins. The SWE decreases for most basins, except the very high Kings basin (73 percent of the basin area is in the upper sub-basin, which has a center of elevation at 2743 m), using the wet and warm Hadley. The peak snowmelt month similarly shifts earlier for the low-elevation basins and is unchanged for the high ones. For the PCM projections, the SWE is significantly reduced, and the peak is earlier for all basins by 2080 through 2099. The critical factor is whether the historical temperature is sufficiently below freezing for the snowpack to be unaffected by a small temperature increase.

In all cases (except the very high-elevation Kings), the SWE decreases as temperature increases. In general, higher elevation basins are less sensitive

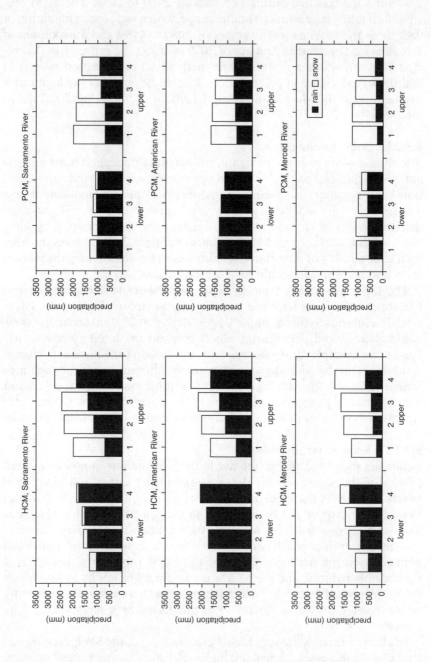

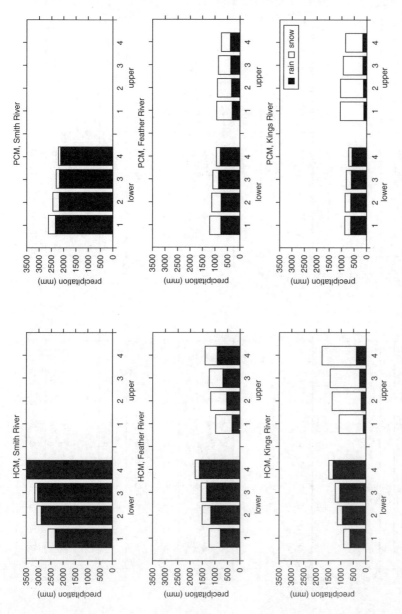

Figure 8.2   Snow (clear) and rain (solid) mean annual depth for the lower and upper sub-basins for (1) each climate baseline, (2) 2010–25, (3) 2050–79, and (4) 2080–99

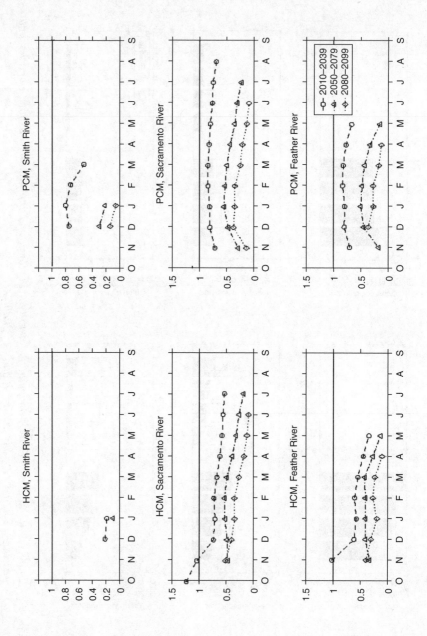

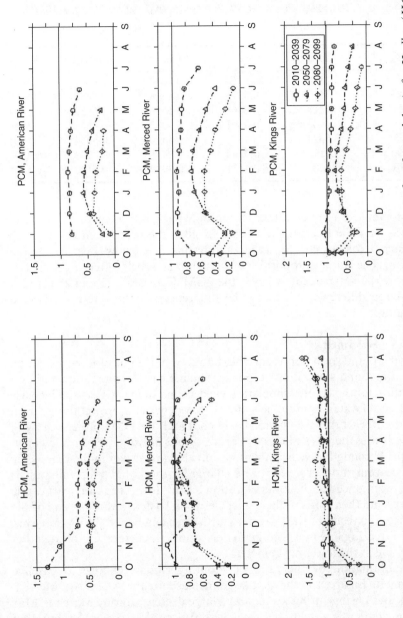

Figure 8.3   Ratio of climate change to baseline mean-monthly snow water equivalent for each basin for Hadley (HCM) and PCM

133

*Table 8.2   Proportion of January 6 h timesteps below freezing for each
upper basin for projected climatological periods 2025
(2010–29), 2065 (2050–79), and 2090 (2080–99) for Hadley
(H) and PCM (P) scenarios*

	Sacramento	Feather	American	Merced	Kings
Baseline	0.7140	0.6710	0.5634	0.6621	0.7002
H 2025	0.5538	0.5215	0.4368	0.5336	0.5619
P 2025	0.7228	0.6661	0.5556	0.6532	0.6895
H 2065	0.4941	0.4591	0.3782	0.4645	0.5014
P 2065	0.5624	0.5336	0.4470	0.5554	0.5901
H 2090	0.3153	0.3164	0.2478	0.3134	0.3546
P 2090	0.5005	0.4731	0.3989	0.5129	0.5449

and do not lose as much winter season snowpack as those with centroid elevations near the freezing line. Table 8.2 gives the proportion of time (6 h timesteps) for which the upper sub-basins are below freezing during January. The Hadley proportion of January that is below freezing decreases by more than 50 percent, whereas the PCM decreases by about 25 percent. The large difference is due to the differences in the rate of projected warming.

8.3.2.3   Snowmelt

Snowmelt and rain represent the liquid water input for evaporation, infiltration, and streamflow response. The increased temperature and precipitation for the Hadley simulation yield a consistent early-season increase in the liquid water input to the hydrologic system as the projections go from 2010 to 2099. Likewise, the relatively cool and dry PCM projection, with temperatures increasing at a slower rate, results in earlier season snowmelt. The peak timing for each simulation shifts toward earlier in the year as the snow-to-rain ratio decreases. The change in the liquid water amount is more pronounced in the lower elevation basins during the early part of the century and then shifts to the higher elevation basins toward the end of the century, as a result of the proximity of the freezing line to the lower basins. As the freezing line moves to higher elevation, the percentage of area that is melting in the lower basin increases.

An evaluation of the ratio of monthly climate change to baseline snowmelt (Figure 8.4) shows a large increase for the American, Merced, and Kings basins during winter and a large decrease during March to May for the Hadley. A similar, but smaller shift occurs for the cooler and drier PCM.

8.3.2.4 Streamflow

The non-linear streamflow response as forced by temperature and precipi-tation change is sensitive to the characteristics of the basin, particularly the snowline elevation and the local weather pattern. Figure 8.5 shows the mean-monthly climatological streamflow for the study basins forced by the two GCM-simulated temperature shifts and precipitation ratios imposed on the historical time series. The warm and wet Hadley-forced streamflow shows large increases in total annual streamflow, increases during the winter and spring seasons (for most of the basins), and earlier peakflow timing for the 2080–99 period. The cool and dry PCM-forced streamflow shows a modest increase in DJF flow volume and decreased JJA streamflow.

The runoff coefficient (streamflow divided by precipitation) increases during November to March and decreases during April to July for the upper sub-basins as forced by both GCM scenarios. This is consistent with the increasing number of days above freezing for each sub-basin.

If temperature and precipitation change uniformly across the state, the streamflow response across the state will still be relatively complex. For the basins studied, a 1.5°C increase is not sufficient for an earlier monthly peakflow. There will, however, be an earlier monthly peakflow at Kings with 3°C warming. Monthly peak flow shifts earlier for the other snow-accumulating sub-basins when warming reaches 5°C. For all the snow-accumulating basins that were evaluated, the December through March monthly streamflow volume increases above the baseline and the May through August volume decreases below the baseline.

8.3.2.5 Cumulative streamflow

The cumulative daily streamflow, starting from the beginning of each water year (1 October), is plotted in Figure 8.6. For both simulations, the day in which 50 percent of the annual flow has occurred is earlier as the projec-tion goes from 2010 to 2100. The Hadley shows very pronounced, large shifts in both the amount and timing; the PCM shows mainly a shift in timing and reduced magnitude. This is consistent with the PCM precipita-tion ratio decreasing. The Hadley streamflow shifts between 30 and 60 days earlier, and the PCM shift is less than or about 30 days near 2100.

8.3.2.6 Exceedence probabilities

Changes in the SWE, coupled with increased wintertime warm precipita-tion (rain), suggest the increased likelihood of more extreme events such as floods. Ranking each set of 30-year peak annual daily flows and generating probability-of-exceedence plots (Figure 8.7) indicates that both the warm and wet Hadley and the cool and dry PCM show a significant increase in the likelihood of high-flow days. For each curve shown, the mean annual

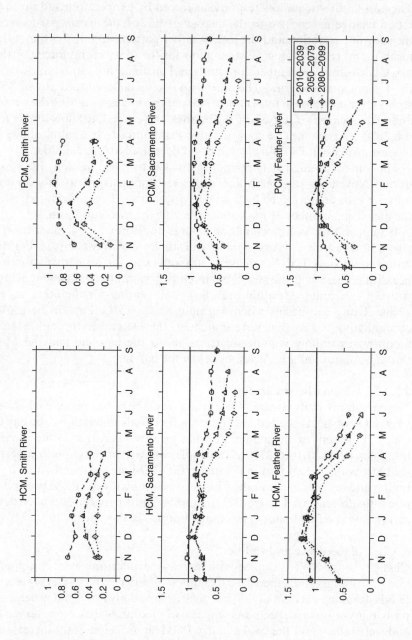

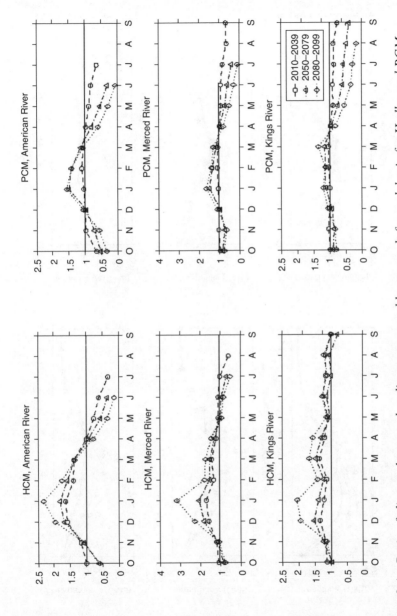

Figure 8.4   Ratio of climate change to baseline mean-monthly snowmelt for each basin for Hadley and PCM

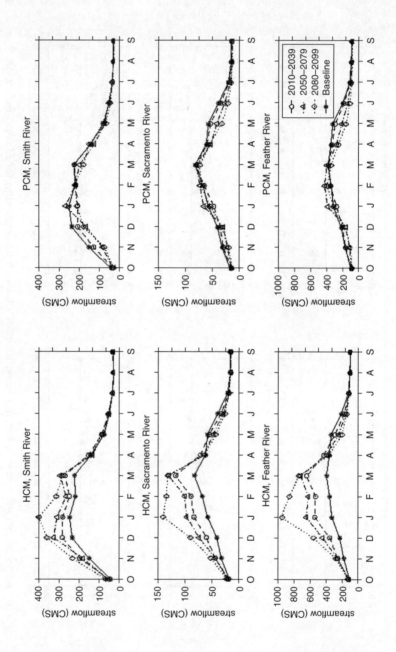

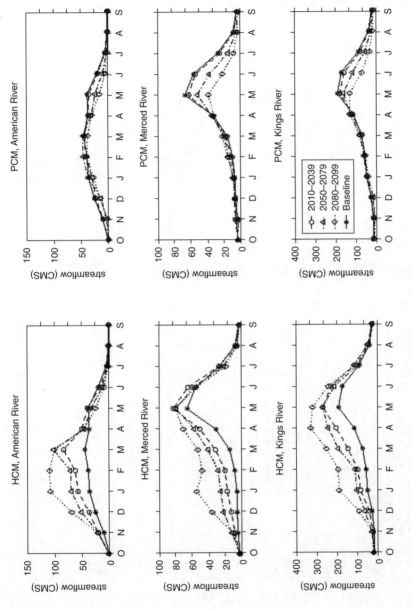

Figure 8.5   Streamflow monthly climatologies based on the Hadley and PCM scenarios

139

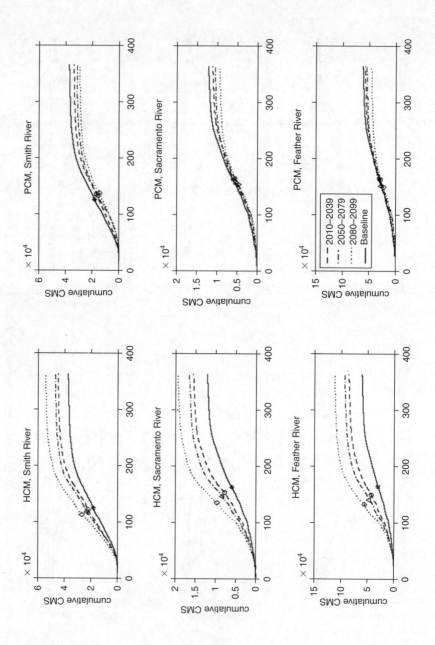

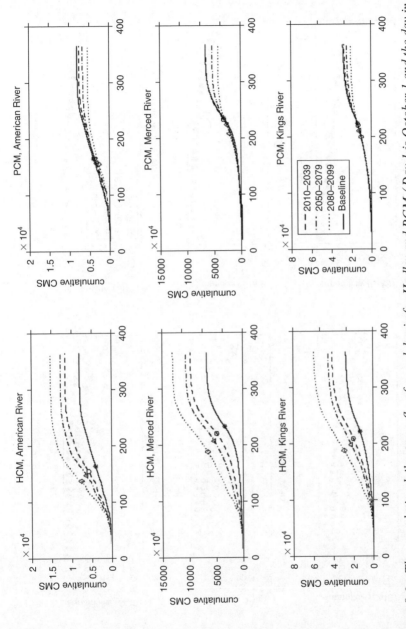

Figure 8.6 The cumulative daily streamflow for each basin for Hadley and PCM (Day 1 is October 1 and the day in which 50 percent of the annual flow has occurred is marked with a symbol)

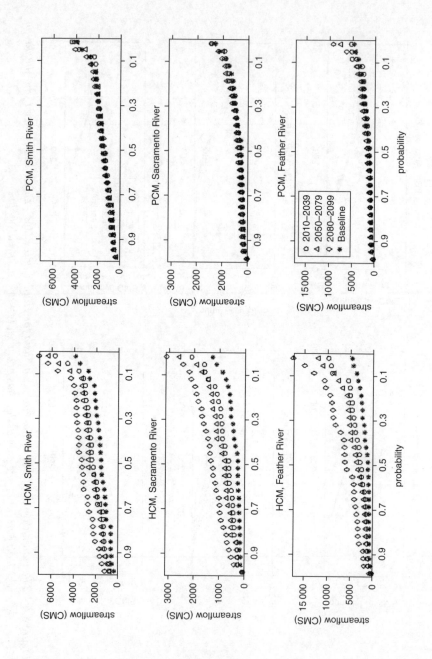

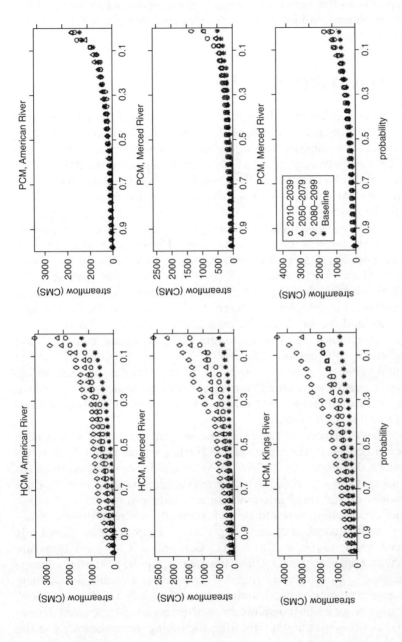

Figure 8.7 Exceedence probabilities of the peak daily flow for each year for each climate change scenario for Hadley and PCM

maximum daily flow is at the 50 percent exceedence interval. Inspecting this figure points to the very large increase in high-flow occurrence and a 5 percent exceedence in high flow for the projected climates that far exceeds current conditions.

8.4   CONCLUSIONS

We analyzed California hydrologic response resulting from temperature shifts and precipitation ratios based on two GCM projections and six specified uniform changes. Streamflow and snowmelt timing shifts are discussed here as lower and upper bounds of the set of possible outcomes. For all cases, with climate change there are fewer freezing days than in the present day during the snowpack storage months. More water flows through the system in the winter and less will be available during the dry season. The large shift in the likelihood of high-flow days is an important result that appears for all cases considered. The results suggest that the range of possible climate change responses is attributable to large-scale change and local characteristics. This could be intensified by large-scale frequency or intensity changes, or both, in natural low-frequency variations (for example the El Niño/Southern Oscillation, the Pacific Decadal Oscillation, and the Arctic Oscillation). Large-scale weather patterns that influence precipitation and runoff timing may shift dynamically, resulting in significantly different local climates.

Monthly changes were superimposed on the historical dataset, so the effects of more intense rainfall events were not represented. The predicted decreasing diurnal temperature ranges (Houghton et al., 2001) were also not represented in this model.

A number of aspects of future climate simulation analysis studies need to be extended. First, there is a need to further evaluate the GCM results and reduce the model bias. More GCM ensemble members of the most recent simulations are needed. An archived sub-daily time series will reduce the amount of statistical interpolation and reduce some errors. Second, dynamic downscaling needs to be incorporated into these studies. A key question is: What scale is of most importance in capturing orographically produced precipitation in California? Certainly GCM resolutions are insufficient, even with the statistical downscaling applied. Another important question is: How many downscaled runs are required and should there be an ensemble of downscaled simulations for each GCM simulation? Third, improving evapotranspiration as a temperature-dependent derivation and channel routing for capturing the timing more accurately in the SAC-SMA is necessary.

Given these limitations, this study does provide an important and reasonable set of upper and lower bounds of hydrologic response to climate change in California. Climate models will never predict the future, but they can yield projections with an uncertainty that can be bracketed. These bracketed solution sets may ultimately give water resource decision makers the type of information needed to safeguard one of our more important natural resources.

ACKNOWLEDGMENTS

We thank the NOAA/NWS California Nevada River Forecast Center for collaborating with us on the application of the SAC-SMA and for providing historical streamflow data. We thank National Center Atmospheric Research and the Hadley Centre for supplying the Arctic Oscillation GCM output. We also thank Ron Nielson and Chris Daley of Oregon State University for furnishing statistically interpolated, 10 km resolution, historical and projected precipitation and temperature time series. Sue Kemball-Cook's help mapping the 10 km fields onto the target watersheds is much appreciated. The project was supported by Stratus Consulting with funds provided to the Electric Power Research Institute by the Commission and by NASA/Regional Earth Sciences Applications Center Grant NS-2791. Work at the US Department of Energy is under contract number DE-AC03-76F00098.

REFERENCES

Anderson, E.A. 1973. National Weather Service River Forecast System: Snow Accumulation and Ablation Model. NOAA Technical Memorandum NWS HYDRO-17. National Oceanic and Atmospheric Administration, Boulder, CO.

Beven, K.J. and M.J. Kirby. 1979. A physically based, variable contributing area model of basin hydrology. *Hydrologic Sciences Bulletin* 24: 43–69.

Burnash, R.J.C., R.L. Ferral, and R.A. McQuire. 1973. *A Generalized Streamflow Simulation System – Conceptual Modeling for Digital Computers.* US Department of Commerce, National Weather Service, and State of California, Department of Water Resources.

Gleick, P.H. 1987. The development and testing of a water balance model for climate impact assessment: Modeling the Sacramento Basin. *Water Resources Research* 23: 1049–61.

Houghton, J.T., Y. Ding, D.J. Griggs, M. Noguer, P.J. van der Linden, D. Xiaosu, and K. Maskell (eds). 2001. *Climate Change 2001: The Scientific Basis.* Cambridge University Press, New York.

Jeton, A.E., M.D. Dettinger, and J.L. Smith. 1996. *Potential Effects of Climate Change on Streamflow: Eastern and Western Slopes of the Sierra Nevada, California and Nevada*. US Geological Survey Water Resources Investigations Report 95-4260. USGS, Washington, DC.

Knowles, N. 2000. Modeling the Hydroclimate of the San Francisco Bay-Delta Estuary and Watershed. PhD Dissertation. Scripps Institution of Oceanography, La Jolla, CA, University of California, San Diego.

Knowles, N. and D.R. Cayan. 2001. Global Climate Change: Potential Effects on the Sacramento-San Joaquin Watershed and the San Francisco Estuary. Available at http://meteora.ucsd.edu/~knowles/papers/GlobalandCalifornia.pdf.

Leavesley, G.H., R.W. Litchy, M.M. Troutman, and L.G. Saindon. 1983. *Precipitation–Runoff Modeling System User's Manual*. US Geological Survey Water Resources Investigations Report 83-4238. USGS, Washington, DC.

Lettenmaier, D.P. and T.Y. Gan. 1990. Hydrologic sensitivities of the Sacramento–San Joaquin River Basin, California, to global warming. *Water Resources Research* **26**: 69–86.

Miller, N.L., J. Kim, R.K. Hartman, and J. Farrara. 1999. Downscaled climate and streamflow study of the southwestern United States. *Journal of the American Water Resources Association* **35**: 1525–37.

Revelle, R.R. and P.E. Waggoner. 1983. Effects of a carbon dioxide induced climatic change on water supplies in the Western United States. In *Changing Climate*. National Academies Press, Washington, DC, pp. 419–32.

Thornthwaite, C.W. and J.R. Mather. 1955. The water balance. In *Publications in Climatology*, Laboratory of Climatology, Drexel Institute of Technology, Centerton, NJ. Available at http://www.udel.edu/Geography/CCR/Abstracts/vol_viii_1.htm.

Wilby, R.L. and M.D. Dettinger. 2000. Streamflow changes in the Sierra Nevada, California, simulated using statistically downscaled general circulation model output. In *Linking Climate Change to Land Surface Change*, S. McLaren and D. Kniven (eds). Kluwer Academic Publishers, New York, pp. 99–121.

9. Changes in crop yields and irrigation demand

Richard M. Adams, JunJie Wu, and Laurie L. Houston

9.1 INTRODUCTION

This chapter presents (1) crop yield response functions that estimate the effect of temperature and precipitation on yields of major crops in California; and (2) evapotranspiration (ET) functions that estimate how water use depends on climate for these same crops. These yield and water use functions are then used to estimate the effect of climate change on future yields and crop water requirements. The crop water requirements are important in any analysis of climate change on Californian agriculture because most crops grown in the state are irrigated. These estimates of yield and water use changes are presented for a range of climate change scenarios and also include assumptions about technological progress and the effects of carbon dioxide (CO_2) levels on crop yields.[1]

The crop yield estimations are an important input for the Statewide Water and Agricultural Production (SWAP) model used in Chapter 11. The information on water requirements is a critical input to the water allocation model in Chapter 10. One of the important innovations in this book is the integration of water and agriculture described in Chapters 10 and 11.

In the past decade, a number of studies have estimated the effects of potential climate change on crop yields (see reviews by Rosenzweig and Hillel, 1998, or Adams et al., 1999a). These studies focused primarily on major grains and oilseed crops, given their importance in total US agricultural production. Some of the more recent investigations explored implications for certain warm season or specialty crops such as fruits or vegetables (for example Adams et al., 1999b). Some studies also estimated changes in ET for irrigated crops. The most complete collection of data on crop yields and water use was generated in support of the recent US National Assessment of Climate Change (Reilly et al., 2001). However, many of the crops grown in California are not included in recent national studies. In addition, many of the crops evaluated in national studies are rain-fed, whereas the majority of California's crops are irrigated. In this study, then, we developed new crop yield and water use estimates appropriate for California.

Previous investigators employed two general techniques to estimate effects of climate change on agricultural yields and ET: (1) crop biophysical simulation models such as CERES or EPIC (see studies reviewed in Rosenzweig and Hillel, 1998); and (2) statistically based models that use cross-sectional data on crop yields and climate variables to estimate how yields differ across climate zones (Mendelsohn et al., 1999; Segerson and Dixon, 1999). The crop simulation models are calibrated to experimental results from agronomists in controlled settings. The cross-sectional results in Segerson and Dixon (1999) and Mendelsohn et al. (1999) are based on county-level data that reflect actual producer decisions. Each technique has certain advantages. For example, crop simulation models allow the analyst to examine a range of weather possibilities, some of which may lie outside the historical record. Crop simulation models also allow inclusion of future changes in CO_2 levels. Elevated CO_2 levels have been shown to have a yield-enhancing effect on most crops; these yield increases may offset some negative effects of warming on crops. Crop simulation models can include farmer adaptations such as changes in planting dates or crop varieties that may offset adverse effects of climate change. However, the crop simulation models tend not to be economically motivated so they may not reflect what farmers actually do. Statistical models do a better job of capturing the adaptations that farmers actually do, but they have other problems. First, they cannot simulate CO_2 fertilization since they do not observe variations in CO_2. Second, the impact of future possible climate outcomes that lie outside observed ranges can only be inferred, they cannot be observed. Third, cross-sectional studies are not done in controlled settings, so there is always the chance that unmeasured variables correlated with climate are confounding the results.

In this study, we relied on panel data analysis using statistical procedures similar to those found in Segerson and Dixon (1999) to obtain estimates of yield and ET changes for California crops. The statistical procedures are discussed in the next section. The results for major crops from each production region in California are then discussed. The chapter ends with a discussion of the overall conclusions.

9.2   METHODS

To reflect the spatial heterogeneity of soil, climate, and production systems in the study region, the state is divided into four major production regions based on climate conditions and cropping systems: (1) the Sacramento Valley and the delta; (2) the San Joaquin Valley and the desert; (3) the Northeast and Mountain region; and (4) the Coast region. Each region comprises multiple counties; the counties represent the cross-sectional component of the data.[2] In general, data are available for each county for 1972 to 2000. Climate and cropping systems are relatively more homogenous within each production region than they are across regions.

9.2.1 Crop Yield Functions

For each production region, the major crops are identified and a yield function is estimated for each crop. The dependent variables of the yield functions are crop yields per acre; the independent variables include seasonal temperatures and precipitation levels, land quality measures, and a time trend variable that reflects technological progress. Specifically, the yield functions are:

$$Y_{it} = \alpha_0 + T_{it}\alpha_1 + T_{it}^2\alpha_2 + P_{it}\beta_1 + P_{it}^2\beta_2 + L_{it}\gamma_1 + L_{it}^2\gamma_2 + \delta t + \varepsilon_{it}, \quad (9.1)$$

where:
i = an index of each county in each region
t = an index of year
Y = crop yield per acre
T = a vector of monthly average maximum daily temperatures during the growing season (°F)
T^2 = a vector of variables in T squared
P = a vector of monthly precipitation during the growing season (inches)
P^2 = a vector of variables in P squared
L = a vector of land quality variables
L^2 = a vector of variables in L squared
ϵ = an error term.

A quadratic functional form is used to estimate the yield functions. This functional form allows climate changes to have a non-monotonic effect on crop yield (that is, a potential increase in yields under warming in cooler locations and a decrease in yields under warming in warmer locations as temperatures increase). The quadratic functional form is also one of the most commonly used functional forms for yield functions in the agronomic literature.

9.2.2   Data and Sources

Yield and acreage data for California crops at the county level are from websites sponsored by the National Agricultural Statistics Service (NASS). One website provided data for 1972 through 2000 for barley, corn grain, corn silage, cotton, dry edible beans, oats, rice, sorghum, sugar beets, and wheat (NASS, 2001b). The other provided data back to only 1980 for alfalfa hay, almonds, avocados, grapes (table, raisin, and wine), olives, Valencia oranges, potatoes, prunes, fresh market tomatoes, processing tomatoes, and walnuts (NASS, 2001a).

Precipitation and temperature weather data for each county in California from 1972 through 2000 are from the Western Regional Climate Center's (WRCC) website (WRCC, 2001). Specifically, we used monthly average data, maximum, and minimum temperatures in degrees Fahrenheit and for monthly total precipitation in inches from this site.

Data on consumptive water use for irrigation are from the US Geological Service (USGS) website (USGS, 2001). Data were available by county for 1985 and 1995.

The time series and cross-sectional data used encompassed yield data for 24 crops in 40 counties, and temperature and precipitation data for 40 counties. Details about the total number of observations for each crop and region are available in Tables 1 and 2 at http://www.energy.ca.gov/reports/2003-10-31_500-03-058CF_A09.PDF.

To account for CO_2 'fertilization' effects on yields, we took data from an earlier study by Adams et al. (1998). This study contained yield estimates derived from plant biophysical simulation models (CERES and EPIC), with and without CO_2 effects. We used these differences in yields (with and without CO_2) to adjust the crop yield estimates given here to account for increased CO_2 levels (to 2100).

In addition, we used the estimates from the same simulation models to adjust the water use demanded by each crop. These adjustments include the effect of CO_2 and a variety of possible different climate scenarios.

9.2.3   Estimation Procedures

Seasonal climate variables are included to reflect the impact of climate variations on crop yields. For corn and other crops grown during the spring and summer periods in California, monthly average maximum daily temperatures and monthly precipitation for March through September are included in the yield functions. For the one crop grown in the winter (winter wheat), monthly average maximum daily temperatures and monthly precipitations for October through the next June are included in the yield function.

Two land quality variables – percentages of high- and medium-quality cropland in a county – are included in all yield functions. High-quality cropland is defined as having land capability classes 1 and 2, and medium-quality cropland is defined as having land capability classes 3 and 4.

Because the yield functions are estimated using pooled time-series and cross-sectional data (that is, across counties and over the period from 1972 to 2000 for most crops) and because adjacent correlated months of weather are included in the analysis, both heteroscedasticity and multicollinearity may arise in the yield functions. In the presence of heteroscedasticity, the least squares estimator no longer has minimum variances among all linear unbiased estimators, but is still unbiased and consistent. The least squares estimator in the presence of multicollinearity remains unbiased and in fact is still the best linear unbiased estimator (BLUE). Because all the standard assumptions are still met, the ordinary least squares (OLS) estimator retains all its desirable properties. The major undesirable consequence of multicollinearity is that the variances of the OLS estimates of the parameters of the collinear variables are inflated. As a result, it is uncertain as to which variable deserves the credit for jointly explained variation in the dependent variable (yields). The clear implication of multicollinearity is that estimated coefficients of any particular climate variable should be interpreted with caution (Segerson and Dixon, 1999). Most important, the result that a particular coefficient is insignificant at some standard level of significance should not be taken as compelling evidence that a particular variable is a statistically irrelevant explanatory variable. In the presence of multicollinearity, the model is most appropriately used to estimate the effect of changes in climate where all temperature variables or all precipitation variables are changed by some common amount or percentage. When used this way, inaccuracies of a particular coefficient are more likely to be offset by the coefficients of other collinear variables, and their joint impact is more likely to be estimated more accurately. Because the yield functions estimated in this study are used to predict the impact of climate changes on crop yields rather than to identify the importance of a particular coefficient on a particular climate variable, they are all estimated using a least squares estimator.

9.2.4   Irrigation Water Use Equation

A simple statistical model is estimated to assess the effect of climate changes on irrigation water use in California. Because of the limited number of observations and the potential multicollinearity problem, we used a linear specification. The dependent variable is the consumptive water use for irrigation per acre. The independent variables include monthly average maximum daily temperatures from March to August and monthly precipitation from May to July.

A single water use model is estimated for the whole study region (California). Water use data are available from the USGS (average water use per acre across all crops) but for only two years (1985 and 1990). Because of the limited number of observations, we were unable to estimate a water use model for each of the four production regions and for each crop, as we did in the case of crop yield functions. We can assume that the general relationships between temperature and precipitation observed from the aggregate data are a reasonable proxy for the ET levels of each crop.

9.3   RESULTS

9.3.1   Estimation of Yields and ET

Selected statistical results from the OLS estimates for each crop and region (from a total of 46 equations) are presented in Table 9.1.[3] For crops with adequate observations (for example corn for grain), regional-specific yield equations are estimated for each region. When the number of observations for a crop is insufficient to estimate a regional-specific yield equation (mostly fruits and vegetables), a single equation is estimated for the state and then adapted to each region where such crops are grown. Regional-specific yield equations are estimated for corn for grain, corn for silage, barley, grain sorghum, dry beans, oats, winter wheat, cotton, cotton (pima), spring wheat, rice, and sugar beets, although not every crop is grown in every region. The results for Sacramento Valley and the delta include yield equations for corn for grain, corn for silage, barley, grain sorghum, dry beans, oats, winter wheat, wheat durum, rice, and sugar beets. The results for San Joaquin Valley and desert include yield functions for corn for grain, corn for silage, barley, grain sorghum, cotton, cotton (pima), dry beans, oats, winter wheat, spring wheat, rice, and sugar beets. The results for the Northeast and Mountain area include yield functions for corn for grain, barley, oats, and winter wheat. The results for the Coast region include yield functions for corn for grain, corn for silage, barley, cotton, dry beans, oats, and winter wheat. For oranges, alfalfa hay, grapes (table, raisin, wine), tomatoes (fresh and processed), almonds, English walnuts, prunes (dried), olives, avocados, and Irish potatoes, a single yield function is estimated for each crop in California. The single equation is then used in regions growing each crop by adapting it to the climatic and soil data for that region.

The results indicate that the weather variables do in fact predict a substantial portion of the variation in yields across time and place. Examining

Table 9.1   Selected yield regression[a]

Crop	Dry beans	Corn	Rice	Barley	Grapes-wine
Region	Coast	San Joaquin	Sacramento	Mountain	CA
Variables					
Intercept	30 640*	−83.3	−56 040	1205*	−19.5
Mar Temp	−385*	−0.97	−648*	−2.6	1.3*
Mar Temp Sq	6.9*	0.01	4.9*	0.0	−0.01*
Apr Temp	654*	0.87	−45	1.1	0.1
Apr Temp Sq	−4.4*	−0.00	0.4	−0.0	−0.00
May Temp	362*	5.56*	737*	−1.5	−1.1*
May Temp Sq	−2.6*	0.03*	−4.5*	0.0	0.01*
Jun Temp	−890*	−4.83*	−515	−9.7*	0.7
Jun Temp Sq	6.0*	0.02*	2.9	0.1*	−0.00
Jul Temp	−524*	9.41*	888	−15.0*	−1.2*
Jul Temp Sq	3.0*	−0.05*	−4.8	−0.1*	0.01*
Aug Temp	610*	−6.10*	1700*	1.3	0.7
Aug Temp Sq	−3.2*	0.03*	−8.9*	−0.0	0.00
Sep Temp	−43	−2.03	−1190*	−5.1	1.4*
Sep Temp Sq	0.1	0.01	−7.0*	0.0	−0.01*
Mar Prec	30*	1.53*	1	−0.6	−0.1
Mar Prec Sq	−1	−0.15	1.1	0.4	0.01
Apr Prec	−260*	0.52	−315*	−2.1	−0.6*
Apr Prec Sq	77*	−0.16	27.4	0.2	0.07*
May Prec	114*	−0.17	165	1.2	−0.1
May Prec Sq	−57*	0.42	−77.4*	−0.1	−0.03
Jun Prec	−157	0.15	311	−7.3*	0.1
Jun Prec Sq	283*	−0.95	−104	1.8*	−0.10
Jul Prec	274	−0.10	1,091	−3.2	0.4
Jul Prec Sq	−2544*	0.27	−1794*	0.7	−0.36
Aug Prec	1706	−2.35	−795*	1.9	−1.0
Aug Prec Sq	−16 021*	0.94	1126*	−0.6	0.71
Sep Prec	−380*	−2.05	229	0.7	0.3
Sep Prec Sq	209*	0.64	−37.7	−0.2	−0.14
Good Land	1273*	−0.28	−1978*	162*	4.7*
Med Land	−68	−0.14	−1197*	132*	11.1*
Time	−14.8*	0.21*	94.0*	0.86*	0.10*
R Squared	0.98	0.43	0.91	0.58	0.56
N	71	172	239	404	329

Note: a. Significant coefficients are starred.

all the yield regressions (including ones not shown), the models fit the data for most crops (see http://www.energy.ca.gov/reports/2003-10-31_500-03-058CF_A09.PDF). The R-squares for all the estimated crop yield equations for Sacramento Valley and the delta vary from 0.33 to 0.93, with half above 0.64. For the crop yield equations in San Joaquin Valley and desert, R-squares vary from 0.24 to 0.85, with all but three above 0.50. The crop yield equations also fit the data well in the Northeast and Mountain area, with all R-squares above 0.60. The performance of crop yield equations varies across crops in the coastal region, with R-squares that vary from 0.22 to 0.99. The R-squares of the yield equations for the 13 fruits and vegetables vary from 0.21 to 0.82, with all but 1 above 0.40. The most poorly performing regressions are for the crops that were estimated across the entire state.

In addition to the overall 'goodness of fit', the estimated coefficients are generally consistent with agronomic expectations. Tests with the coefficients show that for warm-weather adapted crops (for example corn), an increase in temperatures in cooler regions generally increases crop yields. Likewise, for dry-weather adapted crops (for example some fruits and vegetables), an increase in precipitation during the growing season generally reduces crop yields (primarily through adverse effects on fruit set or on harvest quality). However, because of high correlation of some climate variables, coefficients on them may be statistically insignificant in some of the yield equations. As we pointed out earlier, the least squares estimator in the presence of multicollinearity is still the BLUE. But estimated coefficients of any particular climate variable should be interpreted with caution. These types of statistical models are most appropriately used to estimate the effect of changes in climate where all temperature variables or all precipitation variables are changed by some common amount or percentage.

The statistical results for the water use equation are presented in Table 9.2. The model explains 42 percent of variations of irrigation water use across the study region, but the coefficients are generally insignificant. Again, the use of a long series of consecutive months makes the individual coefficients unstable even though the overall results are reasonable. The results suggest that maximum daily temperatures in July have the largest impact on water use. An increase in the average maximum daily temperatures in July of 1°F will increase irrigation water use by 0.14 acre-feet per acre. Precipitation in June and July reduces irrigation water use. An increase in precipitation of 1 inch in each month reduces consumptive water use for irrigation by 0.28 and 0.22 acre-feet per acre, respectively. To assess the effects of changes in climate on water use (in terms of ET), forecasted changes in temperature and precipitation (either from the uniform scenarios or the general circulation models, GCMs) for each of the four regions are used in the water use

Table 9.2   Water use regression[a]

Independent variables	
Intercept	−4.28
Mar Temp	0.03
Apr Temp	−0.07
May Temp	−0.09
June Temp	0.04
Jul Temp	0.14
Aug Temp	−0.04
May Prec	0.11
Jun Prec	−0.28
Jul Prec	−0.22
R Squared	0.42
Adj R Squared	0.32
Observations	61

Note: a. There are no significant coefficients in this regression.

equation. In addition, the effects of elevated CO_2 on irrigation water use in each region are included in the water use estimates. Procedures for this adjustment are described in the next section.

9.3.2 Simulating the Impact of Climate Changes on Crop Yields and Water Use

The estimated yield functions for each crop are used to simulate the effect of changes in climate, as reflected in each scenario, on crop yields in the four major production regions. A three-step procedure is used in the simulations. First, we estimate the baseline yield for each crop in each production region by substituting the means of all variables into the estimated yield functions. These baseline yields are compared with actual yields to evaluate their reasonableness. Second, we estimate crop yields in each production region under each climate change scenario. Finally, we estimate the impact of changes in climate on crop yields by comparing baseline yields with the estimated crop yields under the suite of climate change scenarios.

The changes in yield, measured as a percent changes relative to the base case, was also examined for two uniform climate scenarios: +3°C, +18 percent increase in precipitation and +5°C, 30 percent increase in precipitation. We assume that carbon fertilization will occur with the heightened levels of CO_2. In these scenarios, we also assume technical change will increase at a 0.25 percent rate.[4] As the rate of technology increases, the proportional benefits of CO_2 fertilization may decline if plants begin to approach their photosynthetic capacity. It is not clear how increasing the rate of technological change will affect climate sensitivity. If it is neutral, plants will have the same proportional reductions with climates change, but simply greater yields. If technical change makes plants more sensitive to climate, the harmful impacts of changes in climate will increase in magnitude. If technical change helps plants adapt to warmer or drier climates, however, the effects may be more beneficial. The climate scenarios generally increase crop productivity. However, table grapes and potatoes are estimated to have reduced yields. Further, many crops in the San Joaquin and desert region are estimated to have reduced yields. This region is already very warm and so any further warming appears to reduce crop productivity. In contrast, in the cooler Coast and Mountain regions, crop yields increase with a small amount of warming. Yields of grapes, almonds, and dried prunes are estimated to increase substantially with warming in these scenarios, but a great deal of uncertainty surrounds these estimates.

Tables 9.3, 9.4, 9.5, and 9.6 examine the effect of each GCM climate scenario (PCM and Hadley for 2010, 2060, and 2100) on crop yields in each region. These estimates assume that CO_2 fertilization will occur and that technological change will continue at a 0.25 percent rate per year. The results indicate that selected crops are hurt in the GCM scenarios in each region. In general, the San Joaquin and desert region is the most sensitive to warming as found in the uniform scenarios. The yields of most of the crops in this region decline with higher temperatures. In the other three regions, climate effects are generally beneficial but there are several exceptions. In the Mountain–Northeast and Coast regions, warming hurts all yields of corn silage, barley, rice, sugar beets, winter wheat, oranges, avocados, and potatoes. The yields of the other crops in these two regions improve. In the Sacramento Valley, barley, fresh tomatoes, olives, and potatoes are likely to be hurt by warming. The yields of the other crops grown in this region are estimated to increase with warming.

The direction and magnitude of the intertemporal yield results across the GCM scenarios are more complicated than the static, regional effects reported above. For example, when increased warming is harmful at the outset, the damages (reductions in yields) generally increase over time. However, for some crops where increased warming is initially beneficial in terms of increased yields, continued warming eventually becomes harmful. This is because at higher temperatures yields eventually start to decline as plants exceed the degree or heat unit threshold. In addition, at higher CO_2 concentrations, there are no additional beneficial effects on growth. So the higher CO_2 no longer compensates for adverse effects of a warmer climate.

Table 9.3   Percent change in yields for the Sacramento and delta region
by GCM scenario, with a 0.25 percent technological change
and CO_2 fertilizer effects

	PCM scenario 2010	PCM scenario 2060	PCM scenario 2100	Hadley scenario 2010	Hadley scenario 2060	Hadley scenario 2100
Crop	% change in yield					
Corn grain	3.81	3.46	4.29	8.93	10.65	12.52
Corn silage	5.38	1.78	2.98	1.61	1.93	−1.22
Barley	12.52	−6.86	−6.19	−15.09	−15.81	−16.53
Sorghum	8.80	2.81	0.75	−5.04	−3.74	−1.92
Dry beans	20.70	3.47	4.98	3.76	6.38	4.79
Oats	13.64	3.00	3.86	−10.34	−9.02	−6.54
Rice	17.88	5.56	3.84	1.68	5.89	5.38
Sugar beets	20.81	0.24	3.93	2.41	3.68	10.02
Winter wheat	15.21	2.13	6.57	−4.74	−0.90	1.18
Valencia orange	44.85	11.16	1.79	32.88	9.05	13.59
Hay alfalfa	10.49	6.61	10.62	4.68	8.23	10.19
Grapes (table, raisin)	41.79	10.15	4.34	−21.36	−13.18	−19.36
Grapes (wine)	25.35	9.92	6.25	−1.58	4.85	8.33
Tomatoes (fresh)	24.04	−3.02	−4.42	−0.19	−10.34	−19.04
Tomatoes (processed)	27.4	1.42	0.11	−2.71	1.22	3.58
Almonds	38.4	29.79	40.19	20.69	30.32	37.24
English walnuts	22.35	8.6	4.86	−11.49	−6.21	1.37
Prunes (dried)	29.24	19.42	32.67	13.46	30.12	48.79
Olives	24.37	−0.05	−6.47	−20.01	−21.24	−23.46
Potatoes	4.41	−3.5	−9.45	−5.14	−8.59	−12.28

Research has demonstrated that elevated levels of CO_2 reduce crop ET, primarily through a reduction in stomatal apertures. For example, controlled experiments that measured crop water use (ET) under elevated CO_2 have shown that most crops produce similar or increased yields with less water (for a discussion of the mechanisms leading to this outcome and a review of recent studies, see Rosenzweig and Hillel, 1998).

To account for CO_2 effects on water use in this study, the regression equation that relates water use (ET) to temperature and precipitation (discussed previously) is modified to reflect the increased water use efficiency associated with elevated CO_2 levels. This modification is based on results from crop simulation modeling exercises performed in support of the research

Table 9.4 Percent change in yields for the San Joaquin Valley and desert region by GCM scenario with a 0.25 percent technological change and CO_2 fertilizer effects

	PCM scenario 2010	PCM scenario 2060	PCM scenario 2100	Hadley scenario 2010	Hadley scenario 2060	Hadley scenario 2100
Crop				% change in yield		
Corn grain	3.25	−1.13	−3.67	−3.26	−2.31	−4.87
Corn silage	6.03	3.42	6.3	10.1	8.5	8.36
Barley	13.72	−7.9	−11.86	−15.67	−13.69	−14.43
Sorghum	4.39	−5.38	−4.79	−9.25	−6.76	−7.04
Cotton (pima)	14.18	−0.81	−3.71	13.19	2.25	7.04
Cotton	13.14	−4.42	−6.16	−11.22	−13.51	−14.22
Dry beans	11.79	−13.59	−20.6	−11.41	−7.57	−8.36
Oats	11.2	−10.76	−18.09	−11.94	−14.07	−23.16
Rice	15.81	−2.06	−3.88	−1.29	−0.98	−4.37
Sugar beets	17.76	−4.19	−2.37	−3.23	−8.71	−11.52
Winter wheat	12.28	−3.25	−3.09	−6.11	−4.43	−5.43
Durum wheat	12.35	−3.44	0.14	8.08	5.18	7.92
Valencia orange	27.98	−6.14	−13.63	14.96	5.58	10.27
Hay alfalfa	11.81	7.55	12.07	3.28	5.47	5.75
Grapes (table, raisin)	35.04	−11.84	−5.7	−17.61	−41.6	−38.08
Grapes (wine)	27.55	7.76	6.33	0.61	2.72	6.09
Tomatoes (fresh)	20.61	−9.74	−18.7	−22.76	−32.53	−51.97
Tomatoes (processed)	25.21	−1.63	−1.29	−4.85	−1.03	3.78
Almonds	42.33	32.72	45.46	29.79	27.17	38.54
English walnuts	28.16	8.06	4.22	−9.06	−1.02	0.79
Prunes (dried)	38.62	30.41	46.1	18.02	34.96	52.75
Olives	19.44	−17.92	−18.36	−31.46	−42.35	−48.3
Avocados	24.93	−14.74	−5.65	−12.21	−17.7	−12.67
Potatoes	−0.12	−9.38	−14.91	−7.93	−13.59	−16.93

reported in Adams et al. (1999b). Specifically, the study relies on estimated ET changes that Cynthia Rosenzweig generated for an earlier analysis of national agricultural climate impacts (Adams et al., 1999b). A weighted average adjustment factor is used by crops grown in each region. The CO_2-adjusted irrigation water use values are reported in Table 9.7 for the two GCM scenarios. The Hadley scenario has higher water use than the PCM scenario because it is warmer. Crops have higher water use over time in both GCM scenarios because the temperatures are predicted to increase

Table 9.5 *Percent change in yields for the Northeast and Mountain region by GCM scenario, with a 0.25 percent technological change and CO_2 fertilizer effects*

Crop	PCM scenario 2010	PCM scenario 2060	PCM scenario 2100	Hadley scenario 2010	Hadley scenario 2060	Hadley scenario 2100
			% change in yield			
Corn grain	2.88	3.86	7.76	5.36	4.01	−6.79
Corn silage	−6.53	−8.89	16.88	−3.94	−7.76	−12.2
Barley	14.36	7.08	11.1	7.06	13.17	17.21
Oats	12.74	5.96	2.78	−8.57	−5.3	−3.56
Rice	−10.42	−16.01	−11.64	−26.42	−13.1	−5.98
Sugar beets	9.29	−10.95	−24.15	62.74	86.19	54.97
Winter wheat	10.48	−7.02	−4.98	−12.95	−13.7	−10.7
Hay alfalfa	8.18	11.74	19.63	9.5	14.09	14.57
Grapes (wine)	53.62	40.46	36.73	30.17	36.34	38.08
English walnuts	35.52	31.83	30.57	6.91	16.72	22.25
Potatoes	4.48	−5.3	−7.76	−5.06	−9.44	−11.55

over time. With the uniform climate change scenarios, we find that increased warming and reduced precipitation both increase water use.

In addition to the simulations presented in these tables, we performed a series of sensitivity analyses that included many other cases.[5] Since the end of World War II, national crop yields have increased about 2 percent per year (Huffman and Evenson, 1992). Although there is debate about the sustainability of such yield increases, it is likely that yields will continue to increase in the future. Thus, if a change in climate is predicted for a specific future date, it is likely that future yields will reflect both the climate effects and the impact of technological progress. We thus expanded our technology assumptions to include both a 0 percent and 1 percent annual rate of increase, which bound our base assumption of 0.25 percent/year. In these simulations, the proportional effect of climate change generally declines the higher the rate of technological progress. With high rates of progress, the effect of technology on yields far exceeds the effect of climate on yields. Consequently, the proportional climate impact falls.

In this chapter, we estimated the impact of technological progress on crop yields based on the estimated coefficients on the time trend variable (from the statistical model) and assumptions about the rate of future technological progress. Although experimental results indicate that CO_2 fertilization will occur for most crops, there is the possibility that this effect will be smaller by

Table 9.6   Percent change in yields for the Coast region by GCM scenario, with a 0.25 percent technological change and CO_2 fertilizer effects

	PCM scenario 2010	PCM scenario 2060	PCM scenario 2100	Hadley scenario 2010	Hadley scenario 2060	Hadley scenario 2100
Crop			% change in yield			
Corn grain	11.71	16.83	19.16	33.4	50.98	66.72
Corn silage	−22.25	−29.76	−33.08	−23.32	−31.94	−33.02
Barley	16.71	−1.31	−0.76	−4.05	−2.04	−0.36
Dry beans	21.11	5.83	5.7	9.07	14.98	18.84
Oats	37.17	17.92	16.87	11.36	13.84	22.45
Sugar beets	24.42	10.84	24.25	23.94	31.01	30.17
Winter wheat	−1.12	−8.54	−6.6	−10.42	−5	0.02
Valencia orange	24.91	−7.14	−12.95	13.29	−2.45	−2.2
Hay alfalfa	21.7	14.83	18.33	10.05	13.25	13.37
Grapes (wine)	73.92	39.16	27.7	38.08	29.29	23.65
Tomatoes (fresh)	25.3	9.39	13.04	5.54	4.63	6.16
Tomatoes (processed)	17.85	−6.04	−4.02	−9.54	−4.82	−1.51
Almonds	39.35	27.59	38.11	21.4	23.86	35.06
English walnuts	71.74	40.59	32.91	18.78	18.3	14.64
Prunes (dried)	58.13	40.02	58.42	35.99	41.35	53.31
Avocados	25.87	−2.47	1.03	−3.39	−6.17	−3.24
Potatoes	−10.86	−18.2	−20.96	−16.61	−20.64	−23.28

2100 because of technological changes. We also explored a scenario in which the CO_2 effects are assumed away. When there is no carbon fertilization effect, warming is clearly harmful to crop yields in the San Joaquin and Sacramento valleys. The effects of warming are more mixed in the cooler Coast and Mountain–Northeast regions. The overall impact of warming on crop yields is unequivocally more harmful. A similar result applies to the water requirements of crops. Without CO_2 fertilization, crops will need more water. These results clearly indicate the importance of carbon fertilization in understanding how crops respond to future warming scenarios.

9.4   CONCLUSION

The chapter describes a set of empirical studies that we performed on California crops to determine their response to weather variables. Weather variables explain a large fraction of the observed variation in yields over time

Table 9.7   Percent change in irrigation water use, by GCM scenario accounting for CO_2 effects[a]

Region	PCM scenario 2010	PCM scenario 2060	PCM scenario 2100	Hadley scenario 2010	Hadley scenario 2060	Hadley scenario 2100
	% change in irrigation water use					
Sacramento and delta[b]	−1.6	5.7	17.1	10.0	16.7	19.4
San Joaquin Valley and desert[c]	−5.6	4.3	7.1	9.9	13.8	15.2
Northeast and Mountain[d]	−9.1	0.0	8.8	10.4	15.1	16.3
Coast[e]	−3.4	11.1	18.5	19.4	26.7	31.9

Notes:
a. Temperature and precipitation changes monthly in the GCM scenarios.
b. Sacramento and delta region includes Butte, Colusa, Contra Costa, Glenn, Sacramento, San Joaquin, Solano, Sutter, Tehema, Yolo, and Yuba counties.
c. San Joaquin Valley and desert region includes Fresno, Imperial, Kern, Kings, Madera, Merced, Riverside, Stanislaus, and Tulare counties.
d. Northeast and Mountain region includes Calaveras, El Dorado, Lassen, Maraposa, Modoc, Nevada, Placer, Shasta, Siskiyou, and Tuolumne counties.
e. Coast region includes Lake, Los Angeles, Monterey, Napa, Orange, San Benito, San Diego, San Luis Obispo, Santa Barbara, and Sonoma counties.

and place. Water use by crop was also estimated. The estimated water model was less reliable than the crop yield models because of limited observations. Combining this statistically based information with the results of crop simulation models, we were also able to include the effect of carbon fertilization.

A suite of climate change scenarios was evaluated with the empirical and the simulation model results. Effects of CO_2 fertilization and technological change were both included in the analysis. Because the levels of these latter two variables are not certain, cases with no CO_2 fertilization and higher and lower levels of technical change were also tested. The results depend on these variables, so it is important to use the most appropriate values for these parameters. Without CO_2 fertilization, warming would be more harmful to crops. The greater the rate of technical change, the lower the likely proportional benefit of warming.

A number of general conclusions arise from the results. Warming during the crop-growing season is generally beneficial to yields in the cooler regions of California (the Mountain–Northeast, Coast, and, to a lesser degree, Sacramento–delta regions), but harmful to crop yields in the San Joaquin–desert region. This result is consistent with other studies such as the series of national-level studies (for example Adams et al., 1988, 1995, 1998) that showed gains in crop productivity in more northern latitudes of the United States and losses in some of the southern (and warmer) regions of the country. The explanation for these effects is that crop productivity in cooler regions can benefit from additional degree-days of warming, whereas crops in currently warm regions may already be at the heat threshold level.

A second conclusion is that warming generally increases crop water demand, although the accompanying increases in CO_2 help mitigate these increases. The degree of increase in water use varies across region and climate scenario, but most regions and scenarios show a pattern of increased water demand. Increases in precipitation during the growing season have little impact on water use or crop yields, because virtually all economically important crops in California are grown under irrigated conditions and because rainfall amounts during the state's growing season are typically very limited. Thus, any changes in precipitation are not likely to have a substantial effect on yields or water demand. In fact, for some crops, any increase in precipitation is harmful because precipitation during the growing season may adversely affect crop quality. This is particularly important during the late summer period for many fruit and nut crops.

A third conclusion that emerges from the analysis is the importance of the CO_2 fertilization effect. Specifically, the CO_2 fertilization effect, as used here, increases yield (for those cases where climate change was beneficial) or offsets or mitigates the negative effects of climate change for many crops in regions where changes in climatic variables would otherwise reduce yields. This result is expected, given that yields for each crop in this study were adjusted by a positive factor to reflect the yield-enhancing properties of CO_2. As noted in the methods section, we derived these adjustment factors from previous studies that rely on crop biophysical simulation models. A large number of crop experiments have concluded that elevated CO_2 levels will be beneficial to plant growth (and yields) up to some level of CO_2 increase (here we assume a CO_2 level of approximately 540 ppm in 2060). With higher concentrations of CO_2 and larger magnitudes of climate change, however, it is likely that crop yield effects will tend to be more negative. As noted above, CO_2 is also assumed to decrease crop water use (ET). A fourth conclusion is that rate of technological change is important. As technological change becomes more rapid, the increase in yields over a long time period is dramatic. For example, with a very modest growth rate of 0.25 percent per year, crop yields would increase by almost 30 percent by 2100 over current levels. However, with a 1 percent increase per year, yields would increase almost threefold. Whether such yields are agronomically achievable is open to debate (such increases may exceed the photosynthetic capabilities of the plant). However, technological change will certainly bring about some increases in yields in the next 100 years. By using a base case that includes only a limited level of technological change, the results in the tables in this chapter focus on the climate change effects (the results are similar to the 'climate change only' results). Further, the results presented in this chapter are precisely the cases that are used in Chapter 11.

In summary, this chapter presents a broad range of estimates of the effects of climate change on crop yields and water use. The results should be used to draw general implications about the ability of California's agricultural sector to adapt to climate change. The results show that climate change is not likely to have serious adverse effects on the yields of most California crops. Indeed, most of the results suggest that yields will increase, on average, over current levels, if water supplies remain adequate to produce these crops. Yield increases will allow California to produce current levels of output with fewer acres devoted to agricultural use. In a state facing increased demands for land and water, these yield increases are potentially important. The economic implications of these crop yield and water use results are evaluated in Chapter 10 on water allocation and Chapter 11 on agricultural responses.

NOTES

1. The full report can be accessed at http://www.energy.ca.gov/reports/2003-10-31_500-03-058CF_A09.PDF.
2. Sacramento and delta region includes Butte, Colusa, Contra Costa, Glenn, Sacramento, San Joaquin, Solano, Sutter, Tehema, Yolo, and Yuba counties. San Joaquin Valley and desert region includes Fresno, Imperial, Kern, Kings, Madera, Merced, Riverside, Stanislaus, and Tulare counties. Northeast and Mountain region includes Calaveras, El Dorado, Lassen, Maraposa, Modoc, Nevada, Placer, Shasta, Siskiyou, and Tuolumne counties. Coast region includes Lake, Los Angeles, Monterey, Napa, Orange, San Benito, San Diego, San Luis Obispo, Santa Barbara, and Sonoma counties.
3. The complete set of 46 equations is available in the attachment to Appendix IX at http://www.energy.ca.gov/reports/2003-10-31_500-03-058CF_A09.PDF.
4. Additional climate, technology, and carbon fertilization scenarios are shown at http://www.energy.ca.gov/reports/2003-10-31_500-03-058CF_A09.PDF.
5. See http://www.energy.ca.gov/reports/2003-10-31_500-03-058CF_A09.PDF.

REFERENCES

Adams, R.M., R. Fleming, B. McCarl, and C. Rosenzweig. 1995. A reassessment of the economic effects of global climate change on US agriculture. *Climatic Change* **30**: 147–67.

Adams, R.M., J.D. Glyer, B.A. McCarl, and D.J. Dudeck. 1988. The implications of global climate change for western agriculture. *Western Journal of Agricultural Economics* **13**: 348–56.

Adams, R.M., B. Hurd, S. Lenhart, and N. Leary. 1998. The effects of global warming on world agriculture: An interpretative review. *Journal of Climate Research* **11**: 19–30.

Adams, R.M., B.H. Hurd, and J. Reilly. 1999a. *Agriculture and Global Climate Change: A Review of Impacts to US Agriculture Resources*. Pew Center on Global Climate Change, Arlington, VA.

Adams, R.M., B.A. McCarl, K. Segerson, C. Rosenzweig, K.J. Bryant, B.L. Dixon, R. Conner, R.E. Evenson, and D. Ojima. 1999b. Economic effects of climate change on US agriculture. In *The Impact of Climate Change on the United States Economy*, R. Mendelsohn and J.E. Neumann (eds). Cambridge University Press, Cambridge, UK, pp. 18–54.

Huffman, W.E. and R.E. Evenson. 1992. Contributions of public and private science and technology to US agricultural production. *American Journal of Agricultural Economics* **74**: 751–56.

Mendelsohn, R., W. Nordhaus, and D. Shaw. 1999. The impact of climate variation on US agriculture. In *The Impact of Climate Change on the United States Economy*, R. Mendelsohn and J.E. Neumann (eds). Cambridge University Press, Cambridge, UK, pp. 55–74.

NASS. 2001a. County Agricultural Commissioners' Data. Available at http://www.nass.usda.gov/ca/bul/agcom/indexcac.htm. Accessed September 2001.

NASS. 2001b. US Department of Agriculture National Agricultural Statistics Service: Crops County Data File. Available at http://www.nass.usda.gov/indexcounty.htm. Accessed September 2001.

Reilly, J., F. Tubiello, B. McCarl, and J. Melillo. 2001. The potential consequences of climate variability and change. In *Climate Change Impacts on the United States, Report for the US Global Change Research Program US National Assessment Synthesis Team*. Cambridge University Press, Cambridge, UK, pp. 379–403.

Rosenzweig, C. and D. Hillel. 1998. *Climate Change and The Global Harvest: Potential Impacts of the Greenhouse Effect on Agriculture*. Oxford University Press, New York.

Segerson, K. and B.L. Dixon. 1999. Climate change and agriculture: The role of farmer adaptation. In *The Impact of Climate Change on the United States Economy*, R. Mendelsohn and J.E. Neumann (eds). Cambridge University Press, Cambridge, UK, pp. 75–93.

USGS. 2001. The National Water-Use Program: Downloading water-use information. Available at http://water.usgs.gov/watuse/wudownload.html. Accessed October 2001.

WRCC. 2001. Summary map. Western Regional Climate Center. Available at http://www.wrcc.dri.edu/summary/mapnca.html. Accessed September 2001.

10. Water resources impacts

Jay R. Lund, Tingju Zhu, Stacy K. Tanaka, and Marion W. Jenkins

10.1 INTRODUCTION

In California, concerns about the impact of climate change on water resources have increased in recent years (Dettinger and Cayan, 1995; Gleick and Chalecki, 1999). Several decades of studies have shown that California's hydrology is variable (Cayan et al., 1999) and would change significantly with climate change (Lettenmaier and Gan, 1990; Snyder et al., 2002). The potential effects of climate change on California's water systems have been widely discussed in terms of water availability, flood control, and general water management (Lettenmaier and Sheer, 1991; Gleick and Chalecki, 1999; Wilkinson, 2002).

This chapter focuses on the potential effects of a range of climate change estimates on the long-term performance and management of California's water system. We take a relatively comprehensive approach, looking at the entire intertied California water supply system, including ground and surface waters, agricultural and urban water demands, environmental flows, hydropower, and potential for managing water supply infrastructure to adapt to changes in hydrology caused by climate change. We use an integrated economic-engineering optimization model of California's intertied water system called CALVIN (California Value Integrated Network), which has been developed for general water policy, planning, and operations studies for CALFED (California and Federal Program to manage California's water and aquatic ecosystems) (Jenkins et al., 2001, 2004; Draper et al., 2003). This modeling approach allows us to look at how well the infrastructure of California water could adapt and respond to changes in climate, in the context of higher future populations, changes in land use, and changes in agricultural technology. Unlike conventional simulation modeling approaches, this model economically optimizes the water system to adapt to climate and other changes. It is not limited by present-day water system operating rules and water allocation policies, which by 2100 are likely to be seen as archaic. This approach has its own limitations, but provides useful insights on the potential for operating the current or proposed infrastructure for very different future conditions (Jenkins et al., 2001).[1]

In addition to analyzing statewide water supplies and demand, we also examine the effect of climate change on flooding. With continued urbanization of floodplains and increased spring runoff, future flooding potential and damages could be much worse than today. Therefore, structurally feasible, economically sound, and socially acceptable floodplain management is desirable to reduce vulnerability to damages and balance natural and human uses of floodplains to meet social and economic goals. To illustrate the problems associated with flooding, this chapter combines the influences of urbanization, climate change, and human adaptation in a preliminary case study of flood control problems on California's lower American River near Sacramento, one of the nation's most flood-prone regions.

10.2 METHODS

Many forms of climate change can affect water and water management in California. This chapter examines climate warming and neglects, for the time being, climate variability, sea level rise, and other forms of climate change. Twelve climate change hydrologies are examined to develop integrated statewide hydrologies covering changes in all major inflows to the California water system. For each climate change scenario, permutations of historical flow changes were developed for six representative basins throughout California by researchers at Lawrence Berkeley National Laboratory (see Chapter 8). These changes were used as index basins for 113 inflows to the CALVIN model (Figure 10.1).[2] This comprehensive hydrology includes inflows from mountain streams, groundwater, and local streams, as well as reservoir evaporation for each of the 12 hydrologies. The economic implications of two of these 12 comprehensive changes in California's water availability are then estimated, including effects of forecasted changes in 2100 urban and agricultural water demands.

The CALVIN model explicitly integrates the operation of water facilities, resources, and demands for California's vast intertied water system. It is the first model of California water in which surface waters, groundwater, and water demands are managed simultaneously across the state. The model covers 92 percent of California's population and 88 percent of its irrigated acreage (Figure 10.1), with roughly 1200 spatial elements, including 51 surface reservoirs, 28 groundwater basins, 18 current urban economic demand areas, 24 agricultural economic demand areas, 39 environmental flow locations, 113 surface and groundwater inflows, and numerous conveyance and other links representing the majority of California's water management

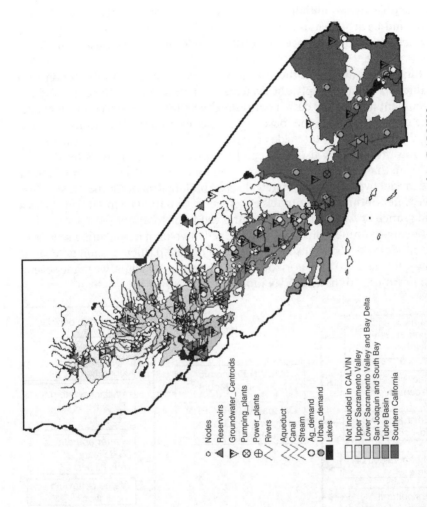

Figure 10.1   Demand areas and major inflows and facilities represented in CALVIN

Legend:

- ○ Nodes
- ◀ Reservoirs
- ▷ Groundwater_Centroids
- ⊗ Pumping_plants
- ⊕ Power_plants
- ∿ Rivers
- ⋙ Aqueduct
- ⋙ Canal
- ⋙ Stream
- ○ Ag_demand
- ◉ Urban_demand
- ■ Lakes

- ☐ Not included in CALVIN
- ☐ Upper Sacramento Valley
- ☐ Lower Sacramento Valley and Bay Delta
- ☐ San Joaquin and South Bay
- ☐ Tulare Basin
- ☐ Southern California

infrastructure. This detailed and extensive model necessitated the assembly and digestion of a wide variety of data within a consistent framework.

The second major aspect of the CALVIN model is that it is an economically driven engineering 'optimization' model. The model, unless otherwise constrained, operates facilities and allocates water to maximize statewide agricultural and urban economic value from water use. This pursuit of economic objectives is initially limited only by water availability, facility capacities, and environmental and flood control restrictions. The model can be further constrained to meet operating or allocation policies, as is done for the base case.[3]

Figure 10.2 illustrates the assembly of a wide variety of relevant data on California's water supply, the systematic organization of the data, and the documentation of the data in large databases for input to a computer code. The model then finds the 'best' water operations and allocations for maximizing regional or statewide economic benefits, and indicates the variety of outputs and their uses that can be gained from the model's results.

More than a million flow, storage, and allocation decisions are made by the model over a 72-year statewide run, making it among the most extensive water optimization models constructed to date. The model produces a wide range of water management and economic outputs.

Owing to limited time and budget, only two climate warming scenarios are modeled explicitly using CALVIN. For this particular climate change study, for the 2100 time horizon with 2100 water demands, we made several modifications to the CALVIN model:

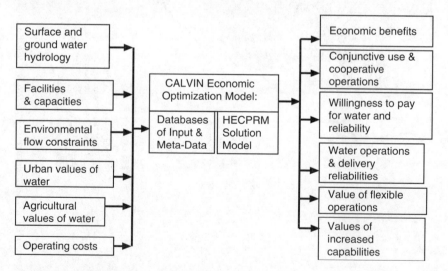

Figure 10.2   Data flow schematic for CALVIN

- Changes in hydrology and water availability for surface and ground-water sources throughout the system to represent different climate change scenarios (Zhu et al., 2005).
- Estimates of 2100 urban and agricultural economic water demands.
- Unlimited access in coastal areas to seawater desalination at a constant unit cost of $1400/acre-ft.
- Urban wastewater reuse available beyond 2020 levels at $1000/acre-ft, up to 50 percent of urban return flows.
- Expansion of local well, pumping, surface water diversion, and connection and treatment facilities to allow access to purely local water bodies at appropriate costs.
- Several corrections to the earlier CALFED version, including revision of environmental requirements.

The method employed for this study contributes several advances over previous efforts to understand the long-term effects of climate change on California's water system, and long-term water management with climate change in general. These include:

- Comprehensive hydrologic effects of climate change, from all major hydrologic inputs, including major streams, groundwater, and local streams, as well as reservoir evaporation. Groundwater, in particular, represents 30–60 percent of California's water deliveries and 17 percent of natural inflows to the system.
- Integrated consideration of groundwater storage. Groundwater contributes well over half of the storage used in California during major droughts.
- Statewide impact assessment. Previous explorations of climate change's implications for California have examined only a few isolated basins or one or two major water projects. However, California has a very integrated and extensive water management system. This system continues to be increasingly integrated in its planning and operations over time. Examination of the ability of this integrated system to respond to climate change is likely to require examination of the entire system. However, the effects of salinization of the Sacramento–San Joaquin delta on freshwater transfers were not examined. In addition, the analysis did not examine changes to or additions to California's water resource infrastructure.
- Economic-engineering perspective. The ability of water sources and a water management system to provide water for environmental, economic, and social purposes, not change in raw water supply, is the relevant measure of the effect of climate change and adaptations to

climate change. Conventional 'yield'-based estimates of climate change effects do not provide results as meaningful as economic indicators of performance.

- Integration of multiple responses. Adaptation to climate change will not be through a single option, but rather a concert of many conventional and new water supply and management options. The CALVIN model explicitly represents and integrates a wide variety of response options.

- Incorporation of future growth and change in water demands. Climate change will have its greatest effects some decades from now. During this time, population growth and other changes in water demands are likely to exert major influences on how water is managed in California and how well this system performs. Scenarios of population and water demand growth were integrated into this analysis.

- Optimization of operations and management. Most climate change impact studies on water management to date have been simulation based. Since major climate changes are likely to occur only after several decades, it seems unreasonable to employ current system operating rules in such studies. Fifty years from now, today's rules and water allocations will be archaic (Johns, 2003). Since water management systems commonly adapt to changing conditions, especially over long time periods, an optimization approach seems more reasonable. Optimization approaches do have limitations (Jenkins et al., 2001), particularly their optimistic view of what can be done. The limitations of optimization seem less burdensome than the limitations of simulation for exploratory analysis of climate change policy and management problems.

The statewide water supply study and results, while preliminary and having obvious and documented limitations (Lund et al., 2003; Tanaka et al., 2006), provide the most comprehensive and detailed study to date of climate warming on California's water system.

Climate change can also have profound effects on flood control and will also interact with other social and economic processes into the future, particularly urbanization. Here, we were only able to examine flooding on one major basin. The floodplain management problem in the American River (see Figure 10.3) is formulated as a long-term optimization problem solved by dynamic programming (DP). While a wide variety of flood management options are available, only levee setbacks and height are considered in this exploratory study. Upstream reservoir control options are not examined explicitly and options to reduce flood damage potential, by building codes,

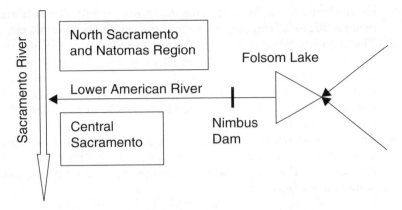

Figure 10.3   Schematic representation of lower American River flood control

zoning, and flood warning and evacuation systems, are similarly neglected for now (Zhu, 2004).

Decision variables in the DP optimization model include (1) levee set-backs on the lower American Rivers and (2) levee heights over long planning periods. As a part of the larger California water system, Folsom Reservoir has several uses other than flood control. To focus on floodplain management, flood control storage decisions are not explicitly included in the DP model. Adaptation of operating rules for the upstream reservoir is an area for further research (Yao and Georgakakos, 2001).

Risk-based optimization is used preliminarily to evaluate economically desirable flood management for the lower American River over a long period of climate change and urbanization. The DP model has the following components and assumptions, which unavoidably simplify the basin's true situation.

10.2.1   Flood Frequency

Flood frequency analysis, as usually practiced, assumes that annual maximum floods conform to a stationary, independent, identically distributed random process. However, with changing flood regimes due to climate change, annual maximum floods in this study are treated dynamically. This dynamic process is represented with three climate scenarios of unimpaired annual peak inflows to Folsom Reservoir.

- The stationary history scenario assumes the peak annual flood distribution is stationary, so the future observes the same lognormal distribution as the historical record (NRC, 1999). The historical mean is 962 m³/s, and the historical standard deviation is 981 m³/s.

- The historical trend scenario assumes continuation of the observed (but perhaps spurious) increasing trend of annual floods, over the planning period. The mean increases linearly at about 5 m³/s each year, and the standard deviation increases at 9 m³/s each year, from the base 2000 level for which the mean is 1200 m³/s and standard deviation is 1338 m³/s.

- The Hadley scenario is represented by trend curves of mean and standard deviation of floods parameterized with hydrological results derived from the Hadley scenario and precipitation runoff studies described in Chapter 8.[4]

Figure 10.4 displays the trends in mean and standard deviation for the stationary history, historical trend, and Hadley scenarios.

In this study, present Folsom Reservoir operating rules are used. Though having some limitations, this simplification allows us to focus on the downstream levee system.

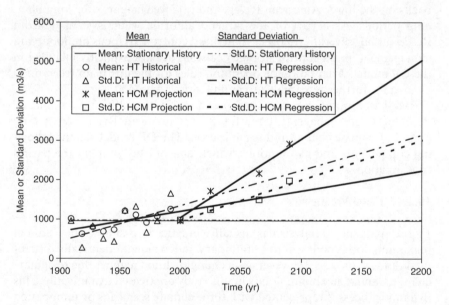

Figure 10.4   Parameterization for mean and standard deviation of three-day floods into Folsom Reservoir for the stationary history scenario, historical trend (HT) scenario, and Hadley (HCM) scenario

10.2.2 Flood Hydraulics

An existing hydraulic model was used to create a set of rating curves for critical locations on the lower American River, for a range of levee setbacks. Hydraulic simulations were made using regulated maximum annual three-day flows.

10.2.3 Damages, Construction Costs, and Land Use Values

The current value of floodplain land protected by the levee is assumed to be $49 422/ha-yr ($20 000/acre-yr); the value of floodplain land by the river is $2471/ha-yr ($1000/acre-yr). Expected flood damage functions are estimated from recent flood damage studies (USACE, 2002). Average damage is estimated to be $11.2 billion at year 2000 level if all land in Sacramento City and Natomas is flooded. Implementation costs of various management decisions are estimated from the literature. The base cost for levee building is estimated to be $35.3/m^3 ($1/ft^3). The lower American River area is rapidly urbanizing. A variety of growth rates for floodplain land values are examined. Damageable property values are assumed to increase at the same rate as floodplain land values. Flood warning systems are already quite good for this area. Thus, potential human life loss due to flood hazard is not employed as part of the objective function. We assume an average discount rate over a long planning horizon, with a base case discount rate of 6.5 percent.

10.3 LIMITATIONS

This study employed an economic-engineering optimization model, which assumes that water is delivered to its highest-valued uses. Among the highest-value uses are urban uses. This generally reflects the likely deliveries of water as well as long-term trends in supply allocation. It is inconceivable that current allocation and operating rules will not evolve over a century. This study assumed there would be no legal impediments to an economically efficient distribution of water supplies; that is, the lags caused by current rules and laws will be much shorter than a century (although they might have very substantial effects for some decades). In reality, while it is reasonable to assume that urban demands for water will be satisfied, legal impediments to transferring water from agriculture to urban or other higher-value uses may remain. The ability to reallocate water as conditions change will not be as smooth or seamless as is assumed. On the other hand, shifts in water allocation and operation are evident even today, and have been substantial over recent decades.

The CALVIN model assumes perfect foresight of year-to-year changes in water resources. This means that water managers know how future supplies will change in advance and can adjust accordingly. Water managers are not omniscient. They must allocate supplies not knowing whether future years will be wet or dry. Allocating water supplies is challenging under current climate because of uncertainties about year-to-year and decade-to-decade variability. This limitation may be particularly significant if climate variability were to increase. The assumption of perfect foresight could significantly understate the economic harm should variability increase. However, the large amounts of over-year storage (particularly in groundwater) and extensive interties make California's system somewhat less vulnerable to such errors.

On the other hand, this study assumes that the current infrastructure remains in place over the twenty-first century with no major modifications. Given how much infrastructure was built in the twentieth century, particularly the latter half of the century, and how much population could increase, there could be major changes to California's water resources infrastructure. In addition, we assume only modest increases in water use efficiencies, and little improvement in desalination, water reuse, and water conservation technologies. These assumptions would tend to overstate damages that can occur.

We examined the implications of flooding for the American River and Sacramento, but we did not study potential damages from increased flooding in the statewide analysis. Chapter 8 notes that peak flows could increase even under the dry PCM scenario. This omission may be important for damage estimates from the wet scenarios, particularly Hadley.

10.4   RESULTS

10.4.1   Changes in Water Demands

An important aspect of future water management is future water demand. California's population continues to grow and its urban areas continue to expand, with likely implications for urban and agricultural water demands. Population growth in California is expected to continue from today's 32 million to 45 million in 2020, with a high population scenario of 92 million by 2100 (see Chapter 2). Overall, water demands could grow significantly (Table 10.1).

The growth in urban population is likely to reduce agricultural land and water use over the next century. The forecast assumes that agricultural land will fall by 750 000 acres by 2100. Nevertheless, there will be an overall

Table 10.1 Change in water demands for California's intertied water system (millions of acre-ft/year)

Use	2020 water	2100 water	2020–2100 change
Urban	11.4	18.6	+7.2
Agricultural	27.8	25.1	−2.7
Total	39.9	44.5	+4.5 maf/yr

Table 10.2 Raw water availability (without operational adaptation, in maf/yr)

Climate scenario	Average annual water availability		Climate scenario	Average annual water availability	
	Volume maf	Change maf (%)		Volume maf	Change maf (%)
1) 1.5T 0%P	35.7	−2.1 (−5.5%)	7) Hadley 2010–39	41.9	4.1 (10.8%)
2) 1.5T 9%P	37.7	−0.1 (−0.4%)	8) Hadley 2050–79	40.5	2.7 (7.2%)
3) 3.0T 0%P	33.7	−4.1 (−10.9%)	9) Hadley 2080–99	42.4	4.6 (12.1%)
4) 3.0T 18%P	37.1	−0.8 (−2.0%)	10) PCM 2010–39	35.7	−2.1 (−5.6%)
5) 5.0T 0%P	31.6	−6.2 (−16.5%)	11) PCM 2050–79	32.9	−4.9 (−13.0%)
6) 5.0T 30%P	36.2	−1.6 (−4.3%)	12) PCM 2080–99	28.5	−9.4 (−24.8%)
Historical	37.8	0.0 (0.0%)			

increase in the demand for water over time. Thus, even without climate change, it is expected that water will be more scarce in the future.

10.4.2 Changes in California's Water Supplies

The 12 climate change scenarios examined, and their overall effects on water availability, appear in Table 10.2. While these are merely raw hydrologic results, adjusted for groundwater storage effects, they indicate a wide range of potential water availability impacts for California's water supply system. These effects range from +4.1 million acre-feet (maf)/yr to –9.4 maf/yr. Figure 10.5 shows the seasonal hydrologic streamflow results for the 12 change scenarios for mountain rim inflows, about 72 percent of California system inflows. For all cases spring snowmelt is greatly decreased with climate warming, and winter flows are generally increased (except for the PCM scenario). These results indicate the overall hydrologic effect of climate change on inflows to California's water supplies. These seasonal changes in runoff have long been identified, based on studies of individual or a few basins (Lettenmaier and Gan, 1990).

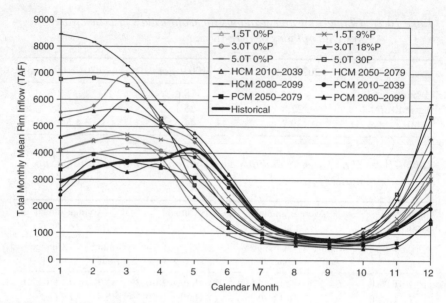

Figure 10.5   Monthly mean rim inflows for the 12 climate scenarios and historical data

10.4.3   Adaptive Changes for Water Management

California has a diverse and complex water management system, with considerable long-term flexibility. Californians are becoming increasingly adept at developing and integrating many diverse water supply and demand management options locally, regionally, and even statewide. The mix of options available to respond to climate change, population growth, and other challenges is likely only to increase in the future with development of water supply and demand management technologies, such as improved wastewater and desalination treatment methods and water use efficiency improvements.

Several statewide scenarios were run using the CALVIN model to evaluate the potential effects of climate warming on California with and without population growth and adaptation. The modeled scenarios included:

- Restricted 2020. This run represents projected water supply operations and allocations in 2020, assuming continuation of current (1997) operation and allocation policies. This run was prepared for CALFED and extensively documented elsewhere (Jenkins et al., 2001, 2004; Draper et al., 2003). This run assumes that inefficient restrictions to urban users continue.

Table 10.3   Summary of statewide operating[a] and scarcity costs

Cost	Restricted 2020	SWM 2020	SWM 2100[b]	PCM 2100[b]	Hadley 2100[b]
Urban scarcity costs	1564	170	785	872	782
Agricultural scarcity costs	32	29	198	1774	180
Operating costs	2581	2580	5918	6065	5681
Total costs	4176	2780	6902	8711	6643

Notes:
a. Operating costs include pumping, treatment, urban water quality, recharge, reuse, desalination, and other variable operating costs for the system. Scarcity costs represent how much users would be willing to pay for desired levels of water delivery.
b. Agricultural scarcity costs are somewhat overestimated because about 2 maf/year of reductions in Central Valley agricultural water demands due to urbanization of agricultural land are not included.

- SWM 2020. This run represents operations, allocations, and perform-ance in 2020 assuming flexible and economically driven operation and allocation policies statewide. This optimized operation can be under-stood as representing operation under a statewide water market, or equivalent economically driven operations. This run also was pre-pared for CALFED and extensively documented elsewhere (Jenkins et al., 2001, 2004; Draper et al., 2003).
- SWM 2100. This run extends the SWM 2020 results to 2100 water demands (Table 10.1), but retains the same (historical) climate used in Baseline 2020.
- PCM 2100. Using the same 2100 water demands as SWM 2100, this run employs the dry and warm PCM 2100 climate change hydrology.
- Hadley 2100. Using the same 2100 water demands as SWM 2100, this run employs the wet and warm Hadley 2100 climate change hydrology.

Population growth will significantly affect the performance and manage-ment of California's vast intertied water system. Overall, population growth alone raises costs by $4.1 billion/yr (the difference between SWM 2020 and SWM 2100 in Table 10.3). Water would be considerably scarcer even without climate change just because of increasing demand. These costs assume that urban users in southern California will be able to pur-chase water from current agricultural users in the Colorado River. If that is not possible, the costs could be considerably higher (see difference between Restricted 2020 and SWM 2020).

Climate change could have large additional effects on this system, espe-cially for the agricultural sector of the economy. These effects are summarized

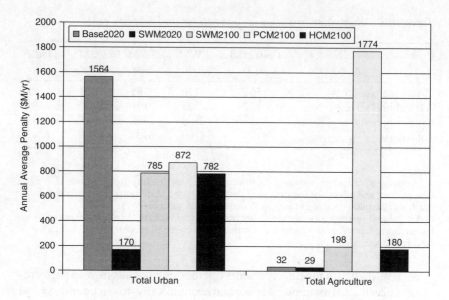

Figure 10.6   Average annual economic scarcity cost by sector

in Table 10.3. Given the 2100 population, these estimates compare the effects of current climate and the two future climates. For example, comparing SWM 2100 with PCM 2100 results reveals that the drier hydrology increases costs by \$1.8 billion/yr. Comparing SWM 2100 with Hadley 2100 results reveals that the wet hydrology decreases total costs by about \$0.25 billion/yr.

Figure 10.6 illustrates how water scarcity effects are shared across sectors. Because urban water demand is more price inelastic, urban use does not change a great deal in either dry or wet scenarios. In contrast, agricultural water use decreases much more if the price of water increases. The economic effects of the drier scenario are most severe for agricultural users.

Not all regions experience the same costs from water scarcity (Figure 10.7). Most water scarcity costs are borne by southern California, with its projected increase in urban water demand. This region does not have enough water to satisfy local demand and must import water from other regions in California and from the Colorado River.

Hydropower production from the major water supply reservoirs in the California system would not be greatly affected by population growth, but would be reduced by the dry PCM 2100 scenario. Base 2020 hydropower revenues average \$161 million/yr from the major water supply reservoirs, compared with \$163 million/yr for SWM 2100. However, the dry PCM

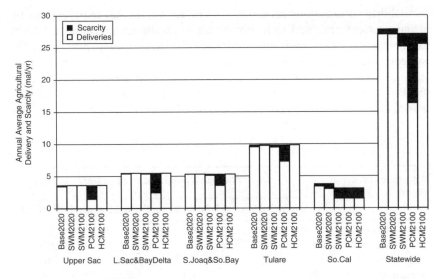

*Figure 10.7   Agricultural water deliveries and scarcity by region and
statewide*

2100 scenario reduces hydropower revenue 30 percent to $112 million/yr.
While this does not include the hydropower impacts of climate change on
other hydropower plants in California, the percentage reduction is probably representative of the overall effect on the state. With the wet Hadley
2100 hydrology, hydropower production greatly exceeds current levels.

CALVIN model results indicate several promising long-term adaptations to population growth and climate change. All 2100 scenarios show
increased market water transfers from agricultural to urban users, additional urban water conservation (~1 maf/yr), use of additional water reuse
treatment (~1.5 maf/yr) and sea water desalination technologies (~0.2
maf/yr), increased conjunctive use of ground and surface waters, and
urbanization of agricultural land. For the dry PCM 2100 scenario, several
million acre-feet/year of reductions in agricultural use occur because of
land fallowing. All of these indicate a much more tightly managed (and
controversial) California water system, where water is increasingly valuable
because water and conveyance capacity are increasingly scarce.

Population growth and climate change also have implications for the
economic costs of environmental constraints. While it appears possible to
comply with environmental flow and delivery requirements for population
growth alone, albeit with some substantial increase in costs to water users,
these costs skyrocket for selected constraints in the PCM 2100 scenario
(see Table 10.4). Increased water demands coupled with decreased water

availability substantially raise the costs of environmental requirements to urban, agricultural, and hydropower users. The increased economic costs of complying with environmental requirements in very dry scenarios suggest that some of these requirements should be amended if California's climate become much drier. The effects of the PCM scenario may be particularly severe for some river basins. For example, under the PCM scenario, there is sometimes not enough runoff in the Trinity and Sacramento

Table 10.4   Shadow costs of selected environmental requirements[a]

Minimum instream flows	Average willingness to pay ($/af)			
	SWM 2020[b]	SWM 2100	PCM 2100	Hadley 2100
Trinity River	0.6	45.4	1010.9	28.9
Clear Creek	0.4	18.7	692.0	15.1
Sacramento River	0.2	1.2	25.3	0.0
Sacramento River at Keswick	0.1	3.9	665.2	3.2
Feather River	0.1	1.6	35.5	0.5
American River	0.0	4.1	42.3	1.0
Mokelumne River	0.1	20.7	332.0	0.0
Calaveras River	0.0	0.0	0.0	0.0
Yuba River	0.0	0.0	1.6	1.0
Stanislaus River	1.1	6.1	64.1	0.0
Tuolumne River	0.5	5.6	55.4	0.0
Merced River	0.7	16.9	70.0	1.2
Mono Lake inflows	819.0	1254.5	1301.0	63.9
Owens Lake dust mitigation	610.4	1019.1	1046.1	2.5
Refuges				
Sac West Refuge	0.3	11.1	231.0	0.1
Sac East Refuge	0.1	0.8	4.4	0.5
Volta refuges	18.6	38.2	310.9	20.6
San Joaquin/ Mendota refuges	14.7	32.6	249.7	10.6
Pixley	24.8	50.6	339.5	12.3
Kern	33.4	57.0	376.9	35.9
Delta outflow	0.1	9.7	228.9	0.0

Notes:
a. Shadow costs are the cost to the economic values of the system (urban, agricultural, hydropower, and operations) of a unit change in a constraint, in this case environmental flow requirements.
b. SWM 2020 results do not include hydropower values (except for Mono and Owens flows).

rivers and the Mono basin to meet current environmental requirements, even with no withdrawals.

10.4.4 Flooding

Flooding effects and adaptations with climate and population changes could not be evaluated for the entire state in this stage of the research. However, a case study of the lower American River provides some sense of the importance of this issue. The DP model was run for a base case and various scenarios. For each comparative analysis, a graph is plotted to demonstrate how parameter perturbations affect the timing and quantities of levee raising and setback changes for the lower American River floodplain. In each graph that shows levee setback and height over a 200-year horizon, the upper half presents levee setback, on the second vertical axis, and the lower half presents levee height, on the primary axis.

Climate change effects
The three climate scenarios (no change, historical trend, and Hadley) are examined first without growth in urban land values. As shown in Figure 10.8, for all climate change scenarios, levee setbacks do not change over the twenty-first century. The stationary history scenario retains the initial

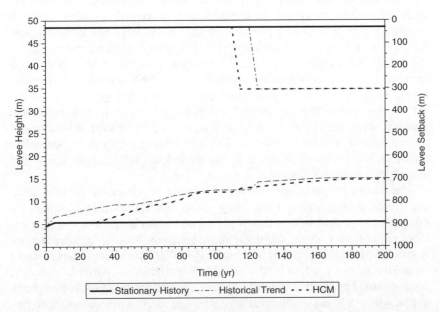

Figure 10.8   Climate change effects on levee setback and height decisions

setback throughout the entire planning horizon while the historical trend and Hadley scenarios expand setbacks in the twenty-second century. These levee setback changes imply that initial setbacks may become economically less than desirable as the climate changes, and that more land should be occupied by the floodway over long periods, even with a tremendous levee relocation cost.

Levee height for the stationary history scenario increases to about 5.4 m at the beginning and has no more change afterward. In contrast, the historical trend and Hadley scenarios have levee height increasing throughout the twenty-first century and then approximately leveling out after that.

Each setback expansion is accompanied by an abrupt levee height increase. Steadily worsening flood frequencies for a high-value urban area lead to economically optimal rising levee heights, punctuated by occasional increases in levee setback.

Combined climate change and urbanization

Without climate change, urbanization alone can drive levee setback change and levee height. Urbanization moves levees closer to the river to increase the protected land and increases levee height to reduce the frequency of flooding for increasingly valuable property. Worsening flood frequency from climate change tends to raise levees and set them back from the river with time. Combining the historical trend, three urbanization rates (0 percent, 2 percent/yr, and 5 percent/yr) and the Hadley climate change scenario, one can see their comprehensive effects on levee construction decisions. Combined with the historical trend climate scenario, the 2 percent/yr urbanization gradually increases levee height to 19.8 m during the twenty-first century. But, with the historical trend scenario, the 2 percent/yr urbanization case expands levee setback twice, but not until the twenty-second century. With the historical trend scenario and 5 percent/yr urbanization, levee height is gradually raised to the same height as with the 2 percent/yr scenario, but sooner. The setbacks are moved in the twenty-first century, also sooner than the 2 percent scenario.

These levee setback and height results have implications for long-term floodplain management. First, they show that the combined effects of climate change and urbanization are more challenging than the impacts from either factor alone, requiring more rigorous flood levee adaptations. Second, with faster urbanization, building higher levees becomes preferable to sacrificing more urban land for floodway expansions. When levee height is constrained by the height limit (19.8 m), the system must resort to setback expansion. At higher urbanization rates, the levee is relocated more frequently after levee height is constrained.

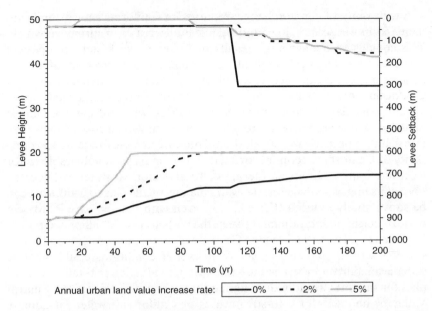

Figure 10.9   Combined effects of Hadley climate change scenario and urbanization

With the Hadley scenario, as shown in Figure 10.9, the levee height of the 2 percent/yr urbanization case gradually increases to the height limit of 19.8 m toward the end of the twenty-first century, with setbacks being increased in the following century.

The 5 percent/yr urbanization case raises the levee height more quickly than does the 2 percent/yr scenario and increases setbacks slightly faster than does the 2 percent/yr scenario. Expected annual flood damages worsen when climate change is combined with urbanization.

10.5   CONCLUSIONS

There are several important conclusions from this chapter. Methodo-logically, it is possible, reasonable, and desirable to include a wider range of hydrologic effects, changes in population and water demands, and changes in system operations in impact and adaptation studies of climate change than has been customary. Including such aspects in climate change studies provides more useful and realistic results for policy, planning, and public education purposes.

A wide range of climate change scenarios for California shows significant increases in wet season flows and significant decreases in spring snowmelt. This conclusion confirms the results of many earlier system component studies. In this chapter, the results are extended to a wider set of hydro-logic changes and extended to calculate how the runoff changes will affect California's major water sources. If done efficiently, California's water system can adapt to the future population growth and climate change modeled, which are fairly severe. This adaptation will be costly in absolute terms, but, if properly managed, should not threaten the fundamental pros-perity of California's economy or society. The single most vulnerable sector to future water scarcity is the agricultural sector. If water is allocated toward its highest-value uses, the water management costs to California can be substantially reduced. For example, even with the very dry PCM sce-nario, efficient adaptation might keep the costs to just $2 billion/year.

Agricultural water users in the Central Valley are the most vulnerable off-stream users to climate change. While wetter conditions could increase water availability for these users, a one-quarter reduction in total water sup-plies for the state would reduce agricultural water deliveries in the Central Valley by about a third. In-stream uses of California's water for aquatic ecosystems and hydropower production are particularly vulnerable to climate change. A reduction in flow from a dry scenario could lead to natural flows that are below current minimum in-stream flow requirements. Dry scenarios may well cause changes in aquatic ecosystems. Changes in precipitation would also affect hydropower production. The dry scenario could reduce hydropower production by 30 percent, whereas the wet sce-nario could increase hydropower 50 percent.

Urban growth in Southern California is projected to raise water demand in this region substantially by the end of the century. Without any add-itional regional sources of water, these urban users will have to import water from current Colorado River agricultural users and possibly from the rest of California. This diversion is limited only by conveyance capacity con-straints on the Colorado River Aqueduct deliveries of Colorado River water and California Aqueduct deliveries of water from the Central Valley. Given the small proportion of local supplies in southern California, the high willingness to pay of urban users for water, and the conveyance-limited nature of water imports, this region is little affected by climate change. Indeed, even in the dry scenario, southern California cannot seek additional water imports. Population growth, conveyance limits on imports, and high economic values lead to high use of wastewater reuse and lesser but sub-stantial use of seawater desalination along the coast.

A case study of the lower American River reveals that flooding prob-lems could be formidable under some wet climate change scenarios. Flood flows indicated by the Hadley 2100 scenario would be well beyond the control capability of existing, proposed, and probably even plausible reservoir capacities. In such cases, major expansions of downstream floodways and changes in floodplain land uses might become desirable. Specifically, there is likely to be economic value to expanding setbacks and levee heights over long periods of time, and making present-day zoning decisions to preserve such options.

While adaptation can be successful overall, the challenges of population growth and climate change are formidable. Even with new technologies for water supply, treatment, and water use efficiency, widespread implementation of water transfers and conjunctive use, coordinated operation of reservoirs, improved flow forecasting, and the close cooperation of local, regional, state, and federal government, the costs will be high and there will be much less 'slack' in the system compared to current operations and expectations. Even with historical hydrology and continued population growth, the economic implications of water management controversies will be greater, motivating greater intensity in water conflicts, unless management institutions can devise more efficient and flexible mechanisms and configurations for managing water in the coming century.

The limitations of this study are considerable, but the qualitative implications seem clear. It behooves us to consider carefully and develop a variety of promising infrastructure, management, and governance options to allow California and other regions to increase the flexibility in the water management system. It is critical that water property rights be defined to facilitate administrative, trading, and market responses to shortages.

Research on adaptation to change for California's water system should be expanded. The analysis should include flood damage costs, sea level rise, salinity concerns, environmental impacts, climate variability, hydrologic refinements, and infrastructure and institutional responses. Other general improvements in the CALVIN model, particularly representations of the Tulare basin, Central Valley groundwater, and agricultural water demands, are also desirable. Climate change studies force us to think bigger and at longer time scales than typically used in our day-to-day professional lives. More comprehensive and adaptive climate change studies are needed to allow scholars and policy makers to focus on actual long-term problems and solutions.

NOTES

1. More detail on the analysis of climate change impacts on supply and demand can be found at http://www.energy.ca.gov/reports/2003-10-31_500-03-058CF_A07.PDF.

2. A color version of the figure can be seen at http://cee.engr.ucdavis.edu/faculty/lund/
    CALVIN/ReportCEC/AppendixA.pdf.
3. The model's detailed schematic and documentation can be found at http://cee.engr.
    ucdavis.edu/faculty/lund/CALVIN/.
4. Climate change is not always gradual. Beyond some threshold, changes can be rapid and
    long-lasting (Alley et al., 2003). An abrupt change version of the Hadley scenario was
    examined, without particularly noteworthy results, and so is not discussed here.

REFERENCES

Alley, R.B., J. Marotzke, W.D. Nordhaus, J.T. Overpeck, D.M. Peteet, R.A. Pielke
    Jr., R.T. Pierrehumbert, P.B. Rhines, T.F. Stocker, L.D. Talley, and J.M. Wallace.
    2003. Abrupt climate change. *Science* **129**: 2005–10.
Cayan, D.R., K.T. Redmond, and L.G. Riddle. 1999. ENSO and hydrologic
    extremes in the western United States. *Journal of Climate* **12**: 2881–93.
Dettinger, M.D. and D.R. Cayan. 1995. Large-scale atmospheric forcing of recent
    trends toward early snowmelt runoff in California. *Journal of Climate* **8**(3):
    606–23.
Draper, A.J., M.W. Jenkins, K.W. Kirby, J.R. Lund, and R.E. Howitt. 2003.
    Economic-engineering optimization for California water management. *Journal
    of Water Resources Planning and Management* **129**(3): 155–64.
Gleick, P.H. and E.L. Chalecki. 1999. The impact of climatic changes for water
    resources of the Colorado and Sacramento-San Joaquin River systems. *Journal
    of the American Water Resources Association* **35**(6): 1429–41.
Jenkins, M.W., A.J. Draper, J.R. Lund, R.E. Howitt, S. Tanaka, R. Ritzema,
    G. Marques, S.M. Msangi, B.D. Newlin, B.J. Van Lienden, M.D. Davis, and
    K.B. Ward. 2001. Improving California Water Management: Optimizing Value
    and Flexibility. Center for Environmental and Water Resources Engineering
    Report No. 01-1. Dept. of Civil and Environmental Engineering, University
    of California, Davis. Available at http://cee.engr.ucdavis.edu/faculty/lund/
    CALVIN/.
Jenkins, M.W., J.R. Lund, R.E. Howitt, A.J. Draper, S.M. Msangi, S.K. Tanaka,
    R.S. Ritzema, and G.F. Marques. 2004. Optimization of California's water
    system: Results and insights. *Journal of Water Resources Planning and Manage-
    ment* **130**(4): 271–80.
Johns, G. 2003. Where is California taking water transfers? *Journal of Water
    Resources Planning and Management* **129**(1): 1–3.
Lettenmaier D.P. and T.Y. Gan. 1990. Hydrologic sensitivity of the
    Sacramento–San Joaquin River Basin, California, to global warming. *Water
    Resources Research* **26**(1): 69–86.
Lettenmaier D.P. and D.P. Sheer. 1991. Climatic sensitivity of California water
    resources. *Journal of Water Resources Planning and Management* **117**(1): 108–25.
Lund, J.R., R.E. Howitt, M.W. Jenkins, T. Zhu, S. Tanaka, M. Pulido, M. Tauber,
    R. Ritzema, and I. Ferriera. 2003. Climate Warming and California's Water
    Future. Center for Environmental and Water Resources Engineering Report No.
    03-1. Dept. of Civil and Environmental Engineering, University of California,
    Davis. Available at http://cee.engr.ucdavis.edu/faculty/lund/CALVIN/.
NRC. 1999. *Improving American River flood frequency analyses.* National Research
    Council Press, Washington, DC.

Snyder, M.A., J.L. Bell, L.C. Sloan, P.B. Duffy, and B. Govindasamy. 2002. Climate responses to a doubling of atmospheric carbon dioxide for a climatically vulnerable region. *Geophysical Research Letters* **29**(11): 10.1029/2001GL014431.

Tanaka, S.K., T. Zhu, J.R. Lund, R.E. Howitt, M.W. Jenkins, M.A. Pulido, M. Tauber, R.S. Ritzema and I.C. Ferreira. 2006. Climate warming and water management adaptation for California. *Climatic Change*, **76**: 361–87.

USACE. 2002. Sacramento and San Joaquin River Basins Comprehensive Study Technical Studies Documentation, Appendix F. Economics Technical Documentation. US Army Corps of Engineers. Available at http://www. compstudy.org/docs/techstudies/appendix_f_economics.pdf. Accessed July 2004.

Wilkinson, R. 2002. *Preparing for a Changing Climate: The Potential Consequences of Climate Variability and Change – California.* A Report of the California Regional Assessment Group for the US Global Change Research Program, Santa Barbara, CA.

Yao, H. and A. Georgakakos. 2001. Assessment of Folsom Lake response to historical and potential future climate scenarios: 2. Reservoir management. *Journal of Hydrology* **249**: 176–96.

Zhu, T. 2004. Climate Change and Water Resource Management: Adaptations for Flood Control and Water Supply. PhD dissertation. Department of Civil and Environmental Engineering, University of California, Davis, CA.

Zhu, T., M.W. Jenkins, and J.R. Lund. 2005. Estimated impacts of climate warming on California water availability under twelve future climate scenarios. *Journal of the American Water Resources Association*, **41**: 1027–38.

11. Agricultural impacts

Richard Howitt and Elizabeth Pienaar

11.1 INTRODUCTION

Because agriculture is widely recognized to be sensitive to climate, the agriculture sector has been the subject of intensive investigations. Agronomists have conducted numerous experiments both in laboratories and in test fields to see how crops would respond to higher temperatures and increased CO_2 (Rosenzweig and Parry, 1994; Reilly et al., 1996). Economists have incorporated these test results into simulation models of the farming sector (for example Adams et al., 1989, 1990, 1999; Easterling et al., 1993; Kaiser et al., 1993). Cross-sectional studies have explored how farm values and net revenues vary across climates (Mendelsohn et al., 1994, 1999). Irrigation has been explored as an adaptation to both drier and hotter conditions (Mendelsohn and Dinar, 2003).

Agricultural water use dominates the California water sector since irrigation uses approximately 82 percent of the existing developed water supply. State agricultural production leads the nation with a farm gate production value of over $20 billion that also generates an annual export value of $11.3 billion. A wide range of high-value perennial crops and low-value annual fodder and grain crops are produced by Californian irrigated agriculture (Johnston and McCalla, 2004). This range of production results in an equally wide range of value and flexibility in water use. For example, irrigation water used by perennial fruit and nut crops has a use value (value of marginal product) of $250-$300 per acre-foot, and this type of irrigation, which comprises about 50 percent of rural water use, has very little ability to adjust to the fluctuations in water supply. On the other hand, an equal area of irrigated agriculture and water quantity is used on lower-value annual crops. Use values in this type of production range from $40 to $80 per acre-foot. In addition, the annual planting cycle of these crops means that given three months' notice, farmers can adjust their annual planting plans to fallow part of the lower-valued crop acreage, and sell the water that these crops would have consumed to farmers with higher-valued less flexible crops, urban water agencies, or environmental water uses (Jenkins et al., 2004).

This chapter goes beyond this initial research to apply the most sophisticated model of water and agriculture yet developed to the study of potential climate change impacts. We use an integrated interdisciplinary approach to analyze the impacts of climate change on California's agricultural industry. The knowledge base for the underlying physical and economic impacts on California agriculture draws from several sources. Changes in water runoff from climate change are drawn from the analysis reported in Chapter 8. The estimation of climate change impacts on crop yield and evapotranspiration comes from the study of crop effects in Chapter 9. The available water for the agricultural sector in future scenarios comes from the California Value Integrated Network (CALVIN) model described in Chapter 10. The change in the availability of agricultural land caused by increased urbanization is developed in Chapter 3.

The net effect of all these changes on the agricultural resource base and relative crop profitability is combined using the Statewide Water and Agricultural Production (SWAP) Model. SWAP takes the projected CALVIN water allocations for regional crop production, the available land, and the climate, and estimates the resulting economic outcomes. SWAP is an economic optimization model that identifies the demand for water for different regions in California, along with the resulting value of agricultural output. By integrating CALVIN and SWAP, one can simultaneously optimize both water allocation and agricultural decisions to determine the optimal response to each climate scenario. The models calculate how optimal outcomes change as climate conditions vary. By comparing these results with baseline conditions, one can measure the welfare impacts of global warming. The comparison also reveals the optimal adaptations to climate that profit-conscious farmers and water managers could make.

SWAP allows for three types of adjustments that farmers can make to mitigate the effects of changes in water and yield. Farmers can adjust the total area of irrigated agriculture in any given region in response to changes in the demand for crops and the supply of both irrigable land and water. Farmers can modify the combination of crops they grow. Again, both physical and economic factors influence these adjustments. For example, if the demand for high-value crops increases, land and water will shift towards these crops, which will, in turn, cause some low-value crops to be reduced or abandoned. The availability of water and changes in yield also affect relative crop proportions because some crops will have greater returns per unit of water than others. Finally, farmers can adjust by substituting among the different inputs used to grow crops. For example, farmers might substitute capital for water. When faced with water shortages during past droughts, farmers invested in more efficient field-level water delivery systems. This investment enabled them to reduce applied water, without reducing crop yield.

In the next section, we describe the SWAP model and how the model is used to investigate climate change. This section includes comments on the assumptions required to make these calculations. Section 11.3 describes the results. It discusses how farmers are likely to adjust to their changing opportunities and what it will mean to their net revenues. The chapter concludes with a discussion of the implications of the results and comments on further research needed.

11.2   METHOD

11.2.1   SWAP Model[1]

The SWAP model is unique in its ability to identify specific agricultural water allocations that are consistent with observed water use and that match the willingness to pay of different agricultural water users for irrigation water supply.[2]

By using a supply–demand approach, SWAP estimates a 'shadow value' per unit of water, by region and month. This approach explicitly recognizes the effect of higher prices on water demand. The objective function used in SWAP maximizes each region's total net returns from agricultural production, subject to the pertinent production and resource constraints on water and land. Production constraints are in the form of functional relationships that describe the productive trade-offs between land and water use efficiency, in conjunction with capital expenditures. The model distributes water supply based on each region's annual water allocation, the local water costs, and the production opportunities facing the region. The model is consistent with microeconomic theory, which asserts that productive decisions are based on marginal conditions.

The model assumes a perfectly competitive market structure in that producers are unable to influence prices in either input or output markets. It follows that each producer is perceived as being relatively small in relation to the market. In some specialized markets, this assumption may not hold. However, for the bulk of Californian goods, it is fair to say that there are many competitors. The model also assumes that water can be redistributed to the highest value users (within the physical limits of current water infrastructure). The model also assumes that water suppliers and users have perfect foresight of upcoming conditions. If a drought lasts for more than one year, the model takes this into account in the first year of the drought. This assumption probably overestimates the degree to which the system can adjust to climate variation.

The model is calibrated against observed data. Published data, however, report average conditions. The divergence between the average and marginal conditions, in the context of either costs or revenues, can be caused by many factors: heterogeneous land, input quality differences, on-farm productive capacity, and economies of scale. Because the farm operators know the true information whereas the model knows only average outcomes, the model may not capture the cropping allocations and technologies used by each farmer. The reliance on average cost data introduces error into the analysis of marginal cost but no known bias.

Although the model has spatial water constraints, which include physical limitations on annual water availability, within a region the optimal solution allows for transfer of water between different months so that the marginal value of water by month and crop is equated. The marginal value of water, the 'shadow value', represents what a buyer would be willing to pay for an additional unit of water. Generally speaking, this additional unit of water would in turn produce additional agricultural output. The value of that output depends on the type of crop grown and the price that is specific to the region. The SWAP model explicitly recognizes each region's unique willingness to pay for water as a function of its productive opportunities and adapts to changing surface supply scenarios.

The model assumes that the shadow price of water is the same across months. This would be the case if monthly water deliveries are sufficiently flexible between months to enable adjustment to changes in the climate or other growing factors. The alternative would be to constrain each month's water allocation to the quantity observed in the base year data. Since we are using a constrained optimization model such monthly constraints would lead to high shadow (marginal) values of water in constrained months and zero shadow values if there was a slight water surplus. These shadow values would lead to switches in the monthly demand functions used in the CALVIN model. Discussion with water district managers confirmed that they have the ability to make shifts in the timing of water deliveries over short periods of a few weeks for quantities that are less than 10 percent of the planned deliveries. To test whether the SWAP model exceeded these limits when run unconstrained by observed monthly deliveries, the model was run with and without the monthly delivery constraints. The difference between the constrained and unconstrained water allocations ranged from 2 percent to 7 percent for all regions, thus confirming that the assumption of efficient water allocation does not result in allocations that depart far from the observed delivery capacity.

Each region has a different production function for each of the crops produced. Within a region, the production of different crops is connected by the restrictions on the total land and water inputs available. Crop production is modeled using a multi-input production model for each region and crop.

The quadratic form of the production function is one of the simplest functional forms that will allow for decreasing marginal returns to add-itional input and substitutability of inputs, as required by theory. Several different agricultural inputs have been aggregated and simplified to aggre-gate measures of land, water, and capital.

Because crop production is a function of land, water, and capital, sub-stitutions among these inputs can take the form of stress irrigation or of substituting capital for applied water. The capital input is an amalgam of labor management and capital used to improve irrigation efficiency under different technologies. By aggregating capital into a general category, the model loses some of the detail of the farmer's choices. However, the approach does not bias the results.

The model, then, captures the three ways in which farmers can adjust crop production when faced with changes in yield, price, or availability of water. The total amount of irrigated land in production can change. This reaction has been observed during California's periodic droughts, when the largest reduction in water use comes from a reduction in irrigated acres. The farmer can change the mix of crops produced so that the value pro-duced by a unit of water is increased. The farmer can substitute inputs. Specifically for California, the farmer can substitute capital for water. The production function is written in general as:

$$y = f(x_1, x_2, x_3) \qquad (11.1)$$

The specific quadratic used in the SWAP model has the form:

$$y = [\alpha_1, \alpha_2, \alpha_3] \begin{bmatrix} x_1 \\ x_2 \\ x_3 \end{bmatrix} - [x_1, x_2, x_3] \begin{bmatrix} \gamma_{11} & \gamma_{12} & \gamma_{13} \\ \gamma_{21} & \gamma_{22} & \gamma_{23} \\ \gamma_{31} & \gamma_{32} & \gamma_{33} \end{bmatrix} \begin{bmatrix} x_1 \\ x_2 \\ x_3 \end{bmatrix}, \qquad (11.2)$$

where y is the total regional output of a given crop, x_i is the quantity of land, water, or capital allocated to regional production of that crop, and α_i and γ_i are estimated coefficients of the model.

We define the total annual quantities of land, water, and capital available in each region as X_1, X_2 and X_3. The total problem defined over G regions and i crops in each region for a single year is:

$$Max \sum_g \sum_i p_i f_{gi}(x_1, x_2, x_3) - \omega_1 x_{i1} - \omega_2 x_{i2} - \omega_3 x_{i3}$$

subject to

$$p_i = d_i \left[\sum_g f_{gi}(\cdot) \right]$$ (11.3)

$$\sum_i x_{i1} \leq X_{1g} \, (Land)$$

$$\sum_i x_{i2} \leq X_{2g} \, (Water)$$

$$\sum_i x_{i3} \leq X_{3g} \, (Capital)$$

where $p_i = d_i[\sum_g f_{gi}(\cdot)]$ is a crop-specific inverse (price dependent) aggregate demand function that determines the price of crop i, and ω_i is the price of input x_i.

SWAP has 21 regions that span the Central Valley of California and three regions in southern California. Figure 11.1 shows the regions and their designations. The crop categories used in SWAP differ between the northern and southern California regions (see Table 11.1). Unfortunately, the data from the two regions are not reported consistently. Because optimization is done for each region independently, the differences in crop categories do not influence the conclusions that result from the SWAP runs.

One of the strengths of the SWAP model is that it is a calibrated non-linear programming model for agriculture. Prices of crops are endogenously determined by the model. The model solves for the equilibrium of supply and demand. Although the model assumes that farmers take prices as constants, it is nevertheless true that prices will change if aggregate supply or demand shifts. Given the importance that California crops have for national and export markets, changes in statewide output levels for certain crops are likely to affect prices. The model allows statewide crop prices to be endogenous, using a demand function for each crop produced in California.

Determining prices is an important and complex question for regional models analyzing climate change. On the one hand, regional production may be a very small fraction of global production and so it will not affect prices. On the other hand, climate change is likely to affect not only regional supply but also global supply. Changes in global supply are likely to cause price changes. As supply increases (decreases), prices will fall (rise). However, because it is likely that agriculture around the world will respond differently to any single climate scenario, it is hard to predict exactly how prices will change. For example, in some scenarios, the supply of a specific product in California may increase but decline elsewhere in the world. In that case, both quantities and prices may go up, which is a best-case scenario for California farmers. Alternatively, the supply of another crop may

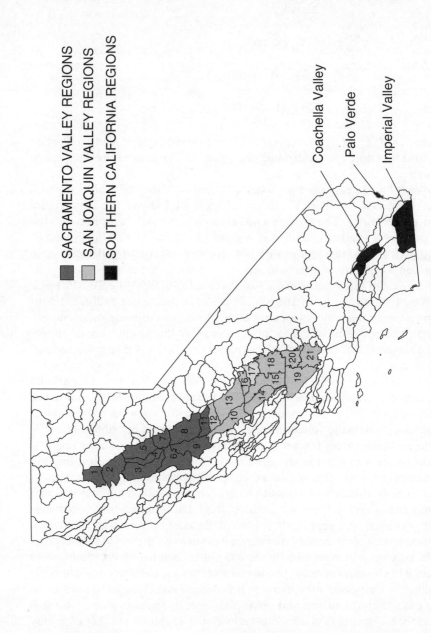

SACRAMENTO VALLEY REGIONS

SAN JOAQUIN VALLEY REGIONS

SOUTHERN CALIFORNIA REGIONS

Coachella Valley

Palo Verde

Imperial Valley

Figure 11.1   California map with southern and northern SWAP regions

Table 11.1   Crop categories used in SWAP

Northern California SWAP crop categories		Southern California SWAP crop categories	
Cotton	Cotton	Cotton	Cotton
Field crops	Field corn	Grain and field crops	Field corn, miscellaneous field crops, and wheat
Fodder	Alfalfa hay, pasture, and miscellaneous grasses	Market crops	Tomatoes and truck crops
Grain crops	Wheat	Low-value crops	Pasture, alfalfa hay, and miscellaneous grasses
Grapes	Table, raisin, and wine grapes	Fruit and nut crops	Orchard and nut crops
Orchard	Almonds, walnuts, prunes, and peaches		
Pasture	Irrigated pasture		
Tomatoes	Fresh market and those for processing		
Rice	Rice		
Sugar beets	Sugar beets		
Subtropical	Olives, figs, and pomegranates		
Truck	Melons, onions, potatoes, and miscellaneous vegetables		

decline in the state but increase in the rest of the world, resulting in falling quantities and prices, which is a worst-case scenario for the state's farmers. Another important question is the magnitude of the price change. An inelastic (steep) demand function will result in large price shifts for relatively small changes in the quantity sold, and the reverse is true for elastic demand functions. California's valuable fruit, nut, and vegetable crops generally have inelastic demand functions because there are a limited number of other producers and consumers have inelastic demand functions for these products. In contrast, the less valuable grains grown in California have more elastic demand functions because many regions grow such crops.

If prices change because of climate change, these price changes will also affect California consumers, not just producers. Increases in prices will make consumers worse off and decreases will make them better off. Because of the relatively large population in the state, these effects on consumers could well be as important as the direct effects of climate on the state's farmers.

11.2.2 SWAP Baseline Scenarios

To prepare SWAP to evaluate climate change impacts, we first had to develop baseline scenarios. Earlier applications of SWAP only looked as far as 2020. To evaluate impacts out to 2100, baseline runs had to be developed far into the future. Climate impacts could then be evaluated relative to these 'business as usual' future scenarios.

Figure 11.2 shows the parameters that were changed and their relationships to other parts of the overall study. First, the model has to make forecasts of changes in the demand for agricultural products, changes in farm technology, and changes in available land and water. We examined parameters needed to match the 'high' population growth scenario in Chapter 2. Population growth affects agriculture because residential users will bid both land and water away from farmers. The model assumes that with the high-growth scenario, population will spill into the Central Valley, displacing marginal farmland.

If urbanization eliminates irrigated acreage in one area, other agricultural lands may shift into irrigation. The model takes into account the loss

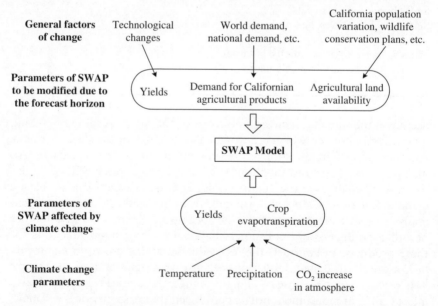

Figure 11.2   Parameters changed in the initial SWAP model

of irrigated lands to urbanization. However, the model does not consider whether new irrigated land might appear as a replacement. This assumption is not expected to be very important.

Given the population growth scenario, the availability of agricultural land in 2100 was then predicted. The amount of land lost to urbanization in each county varied. However, within a county, we assumed that the land lost was equally distributed across the county. We also assumed that the quality of land was homogeneous within each county. These assumptions were required because land use data in SWAP are limited to the percentage of land in each county. In practice, it is not clear what land within a county would be converted to urban uses. Farmers might prefer to offer their less-productive lands for sale, but homeowners might want to buy some of the more fertile lands. The model still preserves variations in productivity between counties. The results for the SWAP regions show that the effects across the state will not be the same and they will depend on regional productivity.

The loss of acres associated with development by 2100 is shown in Figure 11.3 for the north. The decrease in agricultural land is small in the Imperial region and in Coachella Valley. The decrease is larger in Palo Verde, and even larger in the San Diego SWAP region. The largest decreases in land appear to be near Sacramento. However, even in the worst case, the largest loss in any county is only 12 percent of farmland. The decrease in agricultural land is expected to be smaller in percentage terms in the southern area than in the

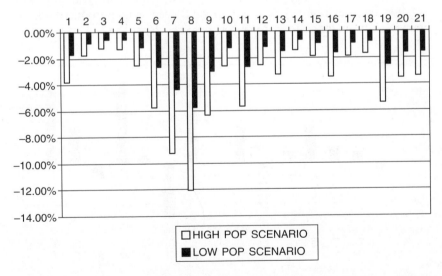

Figure 11.3   Decrease in agricultural land availability from 2020 to 2100 in the northern SWAP regions

northern regions. Overall, the loss of agricultural land to urbanization over the next century is expected to be relatively small.

The model also forecasts the change in crop demands that California would face in 2100 at current real prices. The crop demands in 2100 will be influenced by several factors: competition with emerging or developing countries, trade agreements such as the North American Free Trade Agreement, organizations such as the World Trade Organization, and production in other developed countries. It is not clear how any of these factors will change over the century. To forecast future demand, we use some estimated income elasticities for commodities to predict how demand will shift. Given the forecasted income growth in the United States, we generate changes in crop demands in 2100.

Figure 11.4 shows that the shifts in demand are most important for high-value crops such as tomatoes, market crops, and truck crops (that is, crops that are not processed before selling and directly used or sold fresh, such as fruit, lettuce, celery and flowers). These crops see increases in demand of 100 percent or more for 2100. Orchard crops, grapes, fruits, and nuts see about a 50 percent increase in demand by 2100. The forecasted demand is unchanged for the low-value crops, pasture, and field crops. There is a forecasted decrease in demand for cotton and grain crops.

With the shift in future demand, the model initially predicted the disappearance of the grain crops. However, the grain crops are an important component in many rotational systems. For example, some of the grains are used as fallow crops in a three-year rotation. We consequently introduced a lower bound constraint for grain crop acreage to take such phenomena into account.

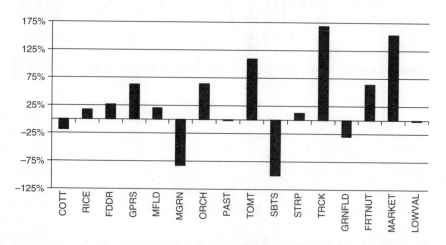

Figure 11.4   Shift in crop demand in California in 2100

The model predicts that by 2100 there will be an increased specialization in high-value crops in California. Indeed, the percentage of land used by orchards, truck crops, tomatoes, and fruit and nut crops may represent nearly 70 percent of the agricultural land available in 2100 (for more detail see http://www.energy.ca.gov/reports/2003-10-31_500-03-058CF_ A10.PDF).

Even choosing a very low technological yield improvement (see Chapter 9), the resulting increases in yields are extremely important. Over the horizon, the 0.25 percent compounded rate of technological change, coupled with the climatic effects, resulted in significant yield increases. Some upper limits for the yields, which are defined on the basis of agronomic potential and differentiated by crop categories, were used to bound the total yield increase and to more conservatively reflect the biological potentialities of the different crops.

The yield changes predicted in Chapter 9 are not uniform across crops and regions. Comparing the different effects of climate change on the yield of orchard versus truck crops reveals that increased summer precipitation may help orchard production by reducing the need for supplemental irrigation, but summer rains will reduce the yield of truck crops that need to be harvested in dry summer conditions.

The regional crop yield changes are introduced into the objective function by multiplying all production functions in each region by a yield increase factor. They were also integrated into the regional market cost because this is a function of the amount of each crop produced.

11.2.3   Climate Change Scenarios

The SWAP model was then run under two different climate scenarios for 2100: Hadley and PCM. The climate scenarios, as discussed in Chapter 8, imply changes in both the amount and timing of runoff. The climate scenarios, as discussed in Chapter 9, imply changes in both the yields and evapotranspiration of crops. These direct effects on plants come from the changes in temperature, precipitation, and CO_2 specified in each climate scenario. As shown in Figure 11.5, the net yield effect of each scenario varies across crops. For example, under the PCM scenario, the productivity of rice, field crops, and pasture all fall. However, the productivity of melon, hay, and orchard crops increases. Under the Hadley scenario, the productivity of many crops is unaffected, but the productivity of cotton and field crops increases.

11.3   RESULTS OF HADLEY AND PCM SCENARIOS

In this section, we compare the results of the baseline to the Hadley and PCM scenarios for 2100 for agriculture in California. The results compare

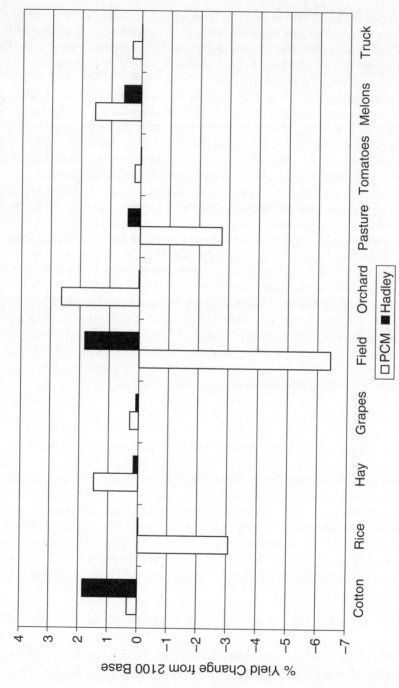

Figure 11.5   Average yield change for PCM and Hadley scenarios

Table 11.2   Fractions of cropland in 2100

	Baseline	Hadley	PCM
Field and rice	11.5	11.4	5.9
High value	70.2	70.1	78.4
Pasture and fodder	11.8	12.1	9.4
Cotton	6.5	6.4	6.3

the baseline result without climate change with the baseline and climate change outcome. The outcome takes into account the change in yield, water supply, water demand, overall land, and crop choice amongst farmers.

Given the amount of land and water available for agriculture, farmers must choose what crop to plant. Table 11.2 compares the cropping pattern in 2100 with and without climate change. With the Hadley scenario, crop choice does not change very much. Farmers grow approximately the same crops as they did in the baseline. However, with the PCM scenario, there is a substantial shift in crop choice. Farmers move away from pasture, field crops, and rice, and move towards the high-value crops (orchards and vegetables). Amongst crops that require water, this reflects a shift from low-valued to high-valued crops.

The climate scenarios also affect the demand for water. That is, they change the amount of water a farmer would want at any specific price. The change in water demand for Region 1 is shown in Figure 11.6 for the baseline, Hadley, and PCM scenarios. The increase in temperature raises the demand for water by each crop significantly. That is, farmers want more water at every price. The effect is especially strong for the PCM scenario because it is so much drier. The demand for agricultural water consequently goes up under global warming. But this effect seems to be shared across all crops and so does not cause a large shift in which crops are grown.

The increase in the demand for water varies by region. For example, the shadow price for water in San Diego (Region 2) can reach $1750/TAF (thousand acre-feet) with a decrease of 25 percent in water availability. In comparison, for other regions, the shadow value of water for a 25 percent decrease in water availability varies from $23 to $600/TAF. The increase in water requirements resulting from climate change in 2100 is much more important for the coastal region (San Diego) than for other SWAP regions. This region has a high concentration of high-valued fruits and nuts. The region consequently would receive a large increase in crop value from more water. Assuming that water is allocated to its highest use, warming would cause more water to be shifted to Region 2.

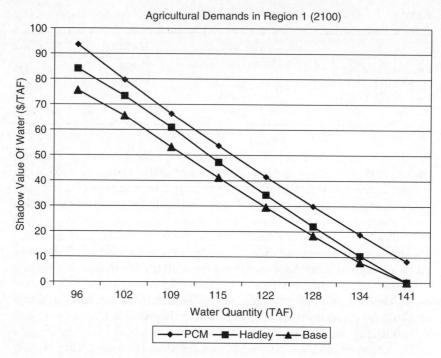

Figure 11.6   Demand for water in Baseline, PCM, and Hadley scenarios

To conduct an integrated analysis between water and agriculture, CALVIN and SWAP were run simultaneously for the baseline, Hadley, and PCM scenarios. CALVIN optimized water allocations for each scenario assuming an efficient statewide water market. That is, CALVIN allocated water to its highest value use. In order to capture interannual fluctuations in runoff, CALVIN relied on a historical 72-year hydrologic record. The socio-economic conditions for 2100 were assumed to apply to each year in the 72-year hydrologic record. The 72-year scenarios for the HadCM2 and PCM scenarios assumed that precipitation in each year changed proportionally in accordance with the average changes predicted by each model outcomes and that temperature increased in absolute terms as predicted by each model. By examining a 72-year pattern, the model was able to explore how agriculture would react across a range of outcomes. One could see how production, profitability, cropping patterns, and water use would change each year. However, this experiment is not perfect because it allows the model to optimize across the actual intertemporal weather pattern rather than keeping future outcomes, such as next year's weather, uncertain.

Table 11.3   *Water allocations in the Hadley and PCM scenarios in 2080–99*

	Overall water quantities and changes					
	Annual		October–March		April–September	
	Quantity (MAF)	Change (%)	Quantity (MAF)	Change (%)	Quantity (MAF)	Change (%)
HadCM2	67.6	78.9	47.5	126.6	20.1	19.3
PCM	28.5	–24.8	17.1	–18.6	11.4	–32.5
Historical	37.8	–	21.0	–	16.8	–

Source: Chapter 10 of this volume.

That is, the model has perfect foresight into what the weather will be next year and takes advantage of this fact, which is not possible in real life. This assumption overstates the adaptability of the system. However, the assumption is applied in both the baseline and the climate scenario. If climate variability does not change, the difference between the baseline and the climate scenario remain the same. But, if variability increases, the perfect foresight assumption understates the resulting damages.

Table 11.3 compares CALVIN's prediction of the water available to agriculture in each climate scenario in 2100. It is clear in Table 11.3 that the Hadley scenario resulted in more water being available while the PCM scenario resulted in less water being available to California farmers. The biggest change in runoff occurs over the winter months when most of the rain in California falls. However, agriculture needs the water during the growing season. The water model, relying on existing dams, moves water across months so that the biggest change in agricultural water used occurs during the growing season.

Figure 11.7 plots the percent difference between the dry PCM and wet Hadley scenarios versus the baseline scenario for four key measures: the percent difference in water used, irrigated land acres, gross revenue, and net income from the crop sector. Note that the percentage change in water availability results in a smaller percentage change in land, and even smaller changes in net revenue. For example, the PCM scenario predicts large reductions in water and irrigated land. However the reduction in gross revenue and especially in net revenue is much smaller. Regions 3, 5, and 7 are projected to lose about 75 percent of their water, 65 percent of their land, but only 40 percent of their net income. Regions 1 and 2 in the Sacramento Valley are projected to lose 50 percent of their water, 35 percent of their

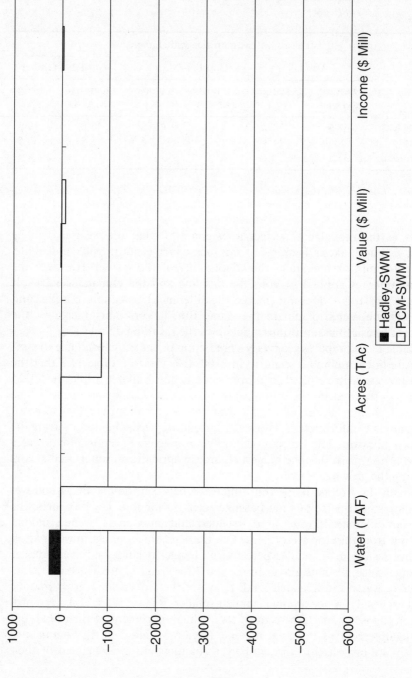

Figure 11.7   Percentage change in water, land, and income from PCM and Hadley scenarios

land, but only 15 percent of their net income. As water is reallocated to its most efficient use and crops change, the impact on net income is limited. Under the Hadley scenario, available water increases. Although this has a minor effect on gross revenue, it has practically no effect on net income. The additional water only compensates for the higher temperature.

The climate scenarios cause water to shift across the four regions. Under the Hadley scenario, the Sacramento and Coastal regions get most of the additional water. Under the PCM scenario, the crops along the coast have the largest increase in irrigation water because they generate the most value per unit of water. Crops in the desert do not get as much of the irrigation water because they yield less value per unit of water. The impact of the PCM scenario falls most heavily in the northern part of the Sacramento Valley. The water model drops water flows to this region precipitously because this part of the Sacramento Valley produces lower-value crops. The value of water in this region is low and so CALVIN takes the water away when overall water is scarce. In contrast, the water supply remains high in other regions with higher-value production. For example, the high-value regions of the San Joaquin Valley see only negligible losses in water and therefore in net income.

The values in Figure 11.7 are average changes over the 72 simulated years. In drought periods, the dry scenarios will translate into substantially larger losses. That is, the effect of the PCM scenario is not uniform every year. The drought-year effects will be worse than the years when there is above-average rain. The average impact shown in Figure 11.7 reflects the fact that droughts will likely be more harmful in the PCM scenario.

In contrast to the PCM scenario, the Hadley scenario generates very slight increases in the economic measures compared to the baseline. Although there is more water available, the higher temperatures have increased water demand. On net, the climate scenarios do not change the economic outcomes in the Sacramento and San Joaquin valleys. Only some districts within these regions are affected. The Yolo, Salano, and Sacramento districts see slight declines in net income with climate change but the Feather River, Tulare Lake Bed, and Fresno districts see large gains. Despite these district fluctuations, the aggregate economic outcomes from the Hadley scenario do not differ significantly from the baseline. Minor increases in regional water supply do not translate into notable differences in any of the economic measures of gross revenue or net income. This is because the higher heat offsets some of the benefits of having more water and also because the additional water supplies would go to relatively low-value crops.

11.4   CONCLUSION

Any prediction about what might happen to California agriculture in 2100 is highly uncertain. Nonetheless, the scenarios in this chapter shed light on what could happen to California agriculture under climate change. The farming system in the state is heavily dependent on irrigation and therefore runoff. Any change in California's water supplies would particularly affect water deliveries to agriculture. However, the water and farming system in the state has an enormous capacity to adjust. Even with relatively dramatic reductions in runoff, for example, the reduction in net farm income will be relatively small.

Because both wetter and drier scenarios are possible for California, we explored both possibilities. The analysis of the wet Hadley scenario shows that despite an increase in runoff, there is only a small increase in net income to California farmers. The increase in water leads to a small increase in low-valued agriculture. In contrast, the dry PCM scenario shows severe cuts in water supply for selected dry periods and regions. The climate scenario results in a 24 percent cut in water supply and a 14 percent reduction in farmland across the state. Despite these large physical changes, however, the SWAP model estimates that net revenues will fall only 6 percent. Most of the losses will knock out low-valued agriculture, preserving the high-valued core of the farming sector. The results suggest a high capacity for California agriculture on the whole to absorb reductions in water supply by focusing those reductions on low-valued agriculture.

The analysis assumes that water will move toward its best use. If institutional barriers prevent water from being reallocated, the damages could be much higher. Further, although average damages are small in the agricultural sector, they are not likely to be the same for every farmer. The model predicts that some regions, which are concentrated in the Sacramento Valley, could have dramatic reductions in agricultural activity and that droughts will cause large damages when they occur. These effects are taken into account in the average calculation, but the average does not describe the range of effects across both space and time.

NOTES

1. This chapter incorporates some parts of Appendix A, Statewide Water and Agricultural Production Model, in the *Improving California Water Management: Optimizing Value and Flexibility* report (Jenkins et al., 2001).
2. More details on this study are provided in http://www.energy.ca.gov/reports/2003-10-31_500-03-058CF_A10.PDF.

REFERENCES

Adams, R., D. Glyer, and B. McCarl. 1989. The economic effects of climate change in US agriculture: A preliminary assessment. In *The Potential Effects of Global Climate Change on the United States: Report to Congress*, D. Tirpak and J. Smith (eds). EPA-230-05-89-050. US Environmental Protection Agency, Washington, DC, pp. 4–1 to 4–56.

Adams, R., B. McCarl, K. Segerson, C. Rosenzweig, K.J. Bryant, B.L. Dixon, R. Conner, R.E. Evenson, and D. Ojima. 1999. The economic effects of climate change on US agriculture. In *The Economic Impact of Climate Change on the Economy of the United States*, R. Mendelsohn and J. Neumann (eds). Cambridge University Press, Cambridge, UK, pp. 18–54.

Adams, R., C. Rosenzweig, R. Pearl, J. Ritchie, B. McCarl, J. Glyer, R. Curry, J. Jones, K. Boote, and L. Allen. 1990. Global climate change and US agriculture. *Nature* **345**: 219–24.

Easterling, W., P. Crosson, N. Rosenberg, M. McKenney, L. Katz, and K. Lemon. 1993. Agricultural impacts of and response to climate change in the Missouri–Iowa–Nebraska–Kansas (MINK) region. *Climatic Change* **24**: 23–61.

Jenkins, M.W., A.J. Draper, J.R. Lund, R.E. Howitt, S. Tanaka, R. Ritzema, G. Marques, S.M. Msangi, B.D. Newlin, B.J. Van Lienden, M.D. Davis, and K.B. Ward. 2001. Improving California Water Management: Optimizing Value and Flexibility. Report Prepared for the CALFED Bay-Delta Program, Sacramento, CA. October. Available at http://cee.engr.ucdavis.edu/faculty/lund/ CALVIN/.

Jenkins, M.W., J.R. Lund, R.E. Howitt, A.J. Draper, S.M. Msangi, S.T. Tanaka, R.S. Ritzema, and G.F. Marques. 2004. Optimization of California's water supply system: Results and insights. *Journal of Water Resources Planning and Management* **130**(3): 271–80.

Johnson, W.E. and A.F. McCalla. 2004. *Whither California Agriculture: Up, Down, or Out? Some Thoughts about the Future*. Special Report 04-1. Giannini Foundation, University of California, Berkeley. August.

Kaiser, H.M., S.J. Riha, D.S. Wilkes, D.G. Rossiter, and R.K. Sampath. 1993. A farm-level analysis of economic and agronomic impacts of gradual warming. *American Journal of Agricultural Economics* **75**: 387–98.

Mendelsohn, R. and A. Dinar. 2003. Climate, water, and agriculture. *Land Economics* **79**: 328–41.

Mendelsohn, R., W. Nordhaus, and D. Shaw. 1994. The impact of global warming on agriculture: A Ricardian analysis. *American Economic Review* **84**: 753–71.

Mendelsohn, R., W. Nordhaus, and D. Shaw. 1999. The impact of climate variation on US Agriculture. In *The Economic Impact of Climate Change on the Economy of the United States*, R. Mendelsohn and J. Neumann (eds). Cambridge University Press, Cambridge, UK, pp. 55–74.

Reilly, J., W. Baethgen, F.E. Chege, S.C. van de Greijn, L. Ferda, A. Iglesias, C. Kenny, D. Patterson, J. Rogasik, R. Rotter, C. Rosenzweig, W. Sombroek, and J. Westbrook. 1996. Agriculture in a changing climate: Impacts and adaptations. In *IPCC (Intergovernmental Panel on Climate Change) Climate Change 1995: Impacts, Adaptations, and Mitigation of Climate Change: Scientific-Technical Analyses*, R. Watson, M. Zinyowera, R. Moss, and D. Dokken (eds). Cambridge University Press, Cambridge, UK, pp. 427–68.

Rosenzweig, C. and M. Parry. 1994. Potential impact of climate change on world food supply. *Nature* **367**: 133–8.

12. Energy impacts

Robert Mendelsohn

12.1 INTRODUCTION

The energy sector plays several roles in climate change policy. Climate change may affect energy supply by changing runoff (as discussed in Chapter 8) and hydroelectricity (as discussed in Chapter 9). The energy sector is also the source of a great deal of greenhouse gases (Metz et al., 2001). Efforts to curtail these emissions will have a direct effect on the energy sector. Finally, climate change is expected to change energy demand as people and firms reduce heating and increase cooling. This chapter focuses strictly on just the third issue, the effect of climate change on energy demand. The analysis examines a national cross-section of energy expenditures to understand how current firms and individuals adapt to being in different climates. The response by energy users to climate is then used to project how future climate change might change energy expenditures in California.

Two methods have been used in the literature to estimate how energy demand might react to climate change: engineering models (Linder et al., 1989; Smith and Tirpak, 1989; Baxter and Calandri, 1992; Rosenthal et al., 1995) and empirical models (Crocker, 1976; Nelson, 1976; Belzer et al., 1996; Morrison and Mendelsohn, 1999; Mendelsohn, 2001). Of course, there are also studies that synthesize earlier research (Nordhaus, 1991; Cline, 1992). The engineering studies calculate the heating or cooling required to maintain a specified temperature given the physical conditions of a site. They can be very precise about specific places with particular characteristics. However, the engineering studies tend not to capture human responses to change. They do not anticipate adaptations in either behavior or physical capital. In contrast, the empirical approaches to measuring the response of households and firms to climate are likely to do a much better job of capturing adaptation. For example, intertemporal studies will show how families and firms actually alter their heating and cooling as outside temperatures change. Cross-sectional studies will not only measure the changes in temperature but also will capture more long-run changes in building characteristics. Empirical studies, however, do not control for unmeasured variables well. There may be some other phenomenon that is correlated with climate that also affects energy. For example, California has an aggressive policy to reduce energy through energy conservation policies. These policies could not be controlled for in this empirical study. Engineering and empirical studies thus have different strengths and weaknesses.

Both engineering and empirical studies suggest that the relationship between energy demand and temperature should be U-shaped – very cool temperatures lead to high heating costs and very warm temperatures lead to high cooling costs. Somewhere between these extreme temperatures there is a temperature that minimizes energy expenditures. The different methods, however, do not always agree about where this minimum lies and how steep the temperature function is. The engineering studies argue that the temperature that minimizes energy expenditures for commercial properties should be cooler than the cost-minimizing temperature for residential buildings. They argue that commercial buildings have more lights that generate excess heat, so their minimum energy expenditure climate should be cooler. However, cross-sectional results reveal that residential buildings actually minimize energy costs in cooler climates than commercial buildings. In practice, homeowners in cooler places do not install cooling capacity (central air-conditioning), so the energy costs of residents are lower in cooler locations. The cross-sectional approach incorporates this human behavior and gets different results.

Energy is an important sector in California. Energy expenditures in California amounted to $71 billion in 2000. The state is a large energy consumer even though energy per capita is low. California accounts for about 11 percent of the nation's electricity consumed even though they have about 16 percent of the population. The state has one of the lowest per capita energy rates in the United States because of aggressive state policies to curb energy use over the last 15 years. Before the program, the state was near the national average. However, at this point, Californian energy use for residential and commercial buildings is about 90 Million British Thermal Units (MBTU)/capita whereas the national average is 124 MBTU/capita. Aggregate residential energy use (52 percent) is slightly larger than commercial use (48 percent).

The demand for energy derives from the demand for many residential and commercial services, including lighting, hot water, refrigeration, and appliance use. However, the part of energy demand most sensitive to climate is space conditioning (heating and cooling). Space conditioning accounts for a large fraction of US residential energy, amounting to 36 percent of electricity demand and 70 percent of natural gas demand in 1990 (Energy Information Administration, 1993). Even in the relatively warm climate of the Pacific region (California, Oregon, and Washington), space

conditioning is an important component of residential energy use accounting for 30 percent of residential energy demand (Energy Information
Administration, 1993).

Climate warming is expected to increase the cost of cooling and decrease
the cost of heating. If firms and individuals try to keep interior temperatures the same and the price of energy remains the same, changes in energy
expenditures will measure the welfare effect of climate change. We test for
the importance of allowing buildings to adjust by estimating short-run and
long-run models. The short-run model holds building characteristics fixed
and the long-run model allows building characteristics to adjust as climate
changes. We review the theoretical basis of this work in the next section and
discuss its limitations.

The results section begins by discussing the energy data used in this study.
The study relied on data from across the United States in order to observe
sufficient climate variation. A two-equation model was estimated using
two-stage least squares. The first stage uses a logit regression to explain the
likelihood that a building is cooled. The second stage estimates energy
expenditures given the predicted cooling value from the first stage. Separate
models were estimated for residential and commercial buildings.

The results section concludes with several forecasts displaying how
climate change might affect California energy demand in the future. A baseline level of energy expenditures is predicted from county-specific growth
in population and per capita income from Chapter 2. This baseline analysis is supplemented by information on building modernization and cooling
penetration from this chapter. Six uniform climate change scenarios are
examined along with the Hadley (Johns et al., 1997) and the PCM (Dai
et al., 2001a, 2001b) scenarios as discussed in Chapter 4. We then predict
the energy impacts (changes in energy expenditures) in each county in
California for each climate scenario using the estimated energy model.

12.2   METHOD

This chapter follows the theoretical structure pioneered by Morrison and
Mendelsohn (1999) with some modification. Climate change is expected to
alter the choices among energy demand and building characteristics that
are derived from the demand for interior temperature. To illustrate this
decision process in the residential and commercial energy sectors, separate
models are developed for households and firms. We take a partial equilibrium approach to this problem and build a model of energy demand independent of other sectors of the economy. Although a general equilibrium
approach would be able to capture interactions among the energy sector

and other climate-sensitive parts of the economy, these interactions are expected to be small, suggesting that such an effort would not be warranted. For example, Mendelsohn and Neumann (1999) estimate that the magnitude of economic impacts on the US economy from climate change is expected to be less than 0.3 percent of gross domestic product (GDP). If impacts are indeed this small, the general equilibrium impacts will be negligible. There is, however, an important interaction between energy impacts and carbon abatement programs that should be considered in future studies, but that is beyond the scope of this work.

12.2.1 Residential Model

Individuals are assumed to choose expenditures for energy, building characteristics, and all other goods to maximize utility subject to a budget constraint. Utility, U, is assumed to be dependent on interior temperature, T, and on an index of all other goods, R. T is assumed to be a function of climate, C, energy use, Q, and building characteristics, Z. The budget constraint exhausts income, Y, on purchases of all other goods, energy, and building characteristics where the price of all other goods is normalized to 1 and P_q and P_z are the prices of energy and building attributes, respectively. The household problem is to choose the level of Q, Z, and R that maximizes U given the budget constraint and climate:

$$\max_{Q,Z,R} U(T, R) \quad s.t. \quad R + P_q \cdot Q + P_z \cdot Z = Y$$

$$T = f(C, Q, Z) \tag{12.1}$$

where:
$$C < T^* (heating) \qquad C > T^* (cooling)$$
$$\Rightarrow T_Q > 0, T_{QQ} < 0 \qquad \Rightarrow T_Q < 0, T_{QQ} > 0$$
$$T_Z > 0, T_{ZZ} < 0 \qquad T_Z < 0, T_{ZZ} > 0$$
$$T_{QZ} > 0, T_C > 0 \qquad T_{QZ} < 0, T_C > 0$$

with T^* representing optimal interior temperature and subscripts representing first and second partial derivatives.

We assume that people must cope with both a heating and a cooling season during a year. During the heating season, the desired interior temperature is assumed to be higher than the outside temperature so that expenditures on both energy (Q) and buildings (Z) increase interior temperatures. During the cooling season, expenditures on energy and buildings reduce interior temperature. An increase in ambient temperature (C) during the heating season consequently reduces expenditures needed to achieve a given interior temperature. An increase in ambient temperature during the cooling season, in contrast, requires more expenditure by the household to maintain initial interior temperatures.

The relationship between energy and building expenditures may be different during the heating and cooling seasons. Building characteristics such as insulation and storm windows tend to reduce energy expenditures in both seasons. Insulating building attributes are substitutes for energy expenditures. If warming reduces the energy expenditures required to keep a home warm, the homeowner may well decide to spend less on expenditures to insulate his home. However, warming will also increase energy expenditures for cooling which should cause home and building owners to increase expenditures on insulation. The net effect is consequently ambiguous. However, some building expenditures are not ambiguous. For example, purchasing cooling equipment only affects summer cooling. Warming will increase the demand for air conditioning, which will unambiguously increase the demand for cooling energy. In this case, one would expect that short-run measures of energy demand that hold building characteristics fixed would have lower climate elasticities than long-run measures of energy demand that allow building characteristics to adjust.

Maximizing equation (12.1) will lead to an optimal bundle of interior temperature, T^*, and all other goods, R^* given initial conditions. With climate change, individuals can respond by changing expenditures on energy, expenditures on building characteristics, or interior comfort levels. We assume that households have a desired interior temperature during the heating and cooling season. If households in the United States are wealthy enough to purchase this desired interior temperature regardless of outside temperatures, the welfare effect of alternative climates simplifies to changes in energy and building expenditures. If warming reduces overall household energy-related expenditures, warming will be a net benefit, but if it increases energy-related expenditures, warming will be a net damage. Note that in many economic analyses, an increase in demand is assumed to be beneficial, not harmful. In this work, however, increases in energy are not assumed to entail increases in utility. Higher energy expenditures imply a higher cost of achieving a specified service, interior temperature, not a higher level of comfort. Increases in cost are damages and decreases in cost are benefits.

The analysis assumes that prices will not be affected by climate change. However, it is likely that warming will increase the demand for electricity in the summer and reduce demand for all fuels in the winter. There should be corresponding price increases for electricity in the summer and price reductions for all fuels in the winter. By assuming prices are constant, the analysis underestimates the losses in summer and underestimates the benefits in winter. However, to the extent that these effects are offsetting over the year, they are expected to be small.

The original research using the energy expenditure approach estimated a single equation model of energy expenditures (Morrison and Mendelsohn, 1999). In this analysis, we develop a two-equation model. In both the residential and commercial models, we begin by estimating the probability that the occupants will cool their building, *Pr(cool)*. The analysis takes a logit functional form:

$$Pr(cool) = exp(\sum BX)/[1 + exp(\sum BX)] \qquad (12.2)$$

where *X* reflects a vector of exogenous variables that might determine cooling such as income, climate, and age of house and *B* is a vector of estimated coefficients.

Earlier cross-sectional studies suggested that homeowners do not cool as much as expected in hot northern summers (Morrison and Mendelsohn, 1999). The results suggested that people endure reduced interior comfort if summers are relatively short. Because Morrison and Mendelsohn (1999) assumed that cooling would be unaffected by warming, they probably underestimated the damages from warming in the summer. In this study, we explicitly predict how cooling will change with warming in the residential sector. The empirical model suggests that warming will encourage people to adopt cooling throughout the residential market. Because this will reduce the current discomfort of warm summers, it will mitigate some of the unmeasured welfare effects in the earlier studies. By including the probability of cooling, the study will do a better job of measuring the welfare effects.

If people maintain interior temperature regardless of climate, one can calculate the response to climate change by fully differentiating the interior temperature production function from (12.1) and setting the sum to zero:

$$T_C d_C + T_Q dQ + T_Z dZ = 0 \qquad (12.3)$$

In the long run, people will change both building and energy expenditures to compensate for a change in climate. In the short run, they will change only energy expenditures. The welfare impacts of climate change on energy can be described as the change in income necessary to pay for the change in energy and building expenditures:

$$dY/dC = P_Q dQ + P_Z dZ \qquad (12.4)$$

If warming increases (decreases) energy expenditures, welfare falls (rises) by the required change in expenditures. Unfortunately, in this study, we cannot measure the cost of building expenditures. All we measure is the change in energy expenditures. By looking at both a short-run and a long-run model, we can determine the importance of building expenditures. If building expenditures are important, there will be a large difference between the short-run and long-run measures.

12.2.2 Firm Model

A parallel model to the household can be constructed for the firm. Instead of maximizing utility, however, the firm maximizes profit. Rather than being constrained by income, the firm is constrained by its production possibility set. This represents the combination of inputs and outputs that are feasible given the production technology. Assuming firms take prices as given exogenously in factor markets, the firm chooses the combination of energy, Q, building characteristics, Z, and all other inputs, X, that minimizes costs subject to its possible production:

$$\min_{Q,Z,X} P_q \cdot Q + P_z \cdot Z + P_x \cdot X \quad \text{s.t.} \quad g(Q, Z, X) = y \quad (12.5)$$

where Q and Z are interior temperature inputs, P_q, P_z, and P_x are respective prices, and y is output. The firm must choose the optimal level of interior temperature and all other inputs that will minimize costs while ensuring production of chosen output Q. Solving equation (12.5) for the optimal combinations of Q, Z, and X yields the well-known economic principle that firms will equate the technical rate of substitution with the economic rate of substitution:

$$\frac{P_q}{P_z} = \frac{g'(Q)}{g'(Z)}. \quad (12.6)$$

Interior temperature measurements suggest that many firms maintain the same interior temperature in both summer and winter across the entire United States. The expenditure necessary to maintain firm profits at their original level is consequently equal to the change in building and energy expenditures needed to maintain interior temperature:

$$dY/dC = P_Q dQ + P_Z dZ \quad \text{s.t.} \quad T_C dC + T_Q dQ + T_Z dZ = 0 \quad (12.7)$$

In this study, the probability of cooling is also estimated for the commercial buildings. Because this probability changes with warming, including it in the model helps address the same biases with respect to changes in interior temperatures as discussed in the residential section. Given this improvement in the model, changes in energy-related expenditures are likely to be more accurate than earlier attempts to use energy expenditures in the commercial sector (Morrison and Mendelsohn, 1999).

12.2.3   Data

This study relies on data from the Energy Information Agency (EIA) of the US Department of Energy. The commercial data comes from the 1989 Commercial Buildings Energy Consumption Survey (Energy Information Administration, 1993) and the residential data from the 1990 Household Energy Consumption and Expenditures Survey (Energy Information Administration, 1992). These surveys provide detailed data on annual energy expenditures and consumption as well as demographics, building characteristics, and climate. The data record choices made for several thousand buildings distributed in random clusters across the continental United States. This sample is weighted to represent the true population of buildings. These weights make it possible later in the analysis to extrapolate from the sample to the population as a whole. One important weakness of this information is that it reports only a single year of expenditures. The energy data reflect the weather of a single year. Ideally, we would want a longer time series of data to capture climate effects.

Although a Californian data set would have the advantage of reflecting unmeasured but unique features of Californians, there are several reasons why we turned to the national scale. First, there is no way to identify California residents in this data set so this was not an option. However, we could identify households from the Pacific region (California, Oregon, and Washington). Unfortunately, there was not enough climate variation in the regional data set to estimate reliable climate coefficients. The study consequently used energy use observations from the entire continental United States. One obvious flaw in the national data set is that energy conservation programs unique to California have not been taken into account.

This study relies on special data that the EIA created for Morrison and Mendelsohn (1999). The public energy survey data were matched with climate data by county from Mendelsohn et al. (1994). Each observation has detailed climate data that are not available in the public survey file. The climate data are long-term (30-year) measurements of temperature and precipitation in each of two months: January and July. The reliance on climate data and not annual weather is an important distinction. Obviously, owners adjust energy expenditures to annual conditions, spending more on energy in cold winters and hot summers. However, because weather changes from year to year, owners do not customize their buildings to endure annual weather. All long-run adjustments are more likely to be keyed to climate, or long-term weather. The results in this study are thus expected to be different from cross-sectional energy studies that have looked only at annual weather (such as Baxter and Calandri, 1992, or Belzer et al., 1996).

The study relies on total energy expenditures by firms and households. This differs from several cross-sectional studies of electricity demand (for example, Baxter and Calandri, 1992, and Sailor, 2001). The electricity studies have certain advantages in that they explain the demand for a specific fuel. However, electricity studies alone will overestimate the damages from warming because electricity has a disproportionate share of cooling costs but a smaller share of heating costs. Electricity studies therefore capture the damages from more cooling but they do not fully capture the benefits from less heating.

The reliance on county measures of climate is not perfect. To protect confidentiality, the EIA added a small error term to the climate data. There is an additional error term from using county-based climate measurements. Some counties are quite large, particularly in California, and actually have a range of climates within their borders.

Another limitation of the data is that energy is not divided between energy needed for temperature control and energy needed for other purposes. We assume in this analysis that the energy households and firms use for other purposes is not correlated with climate. Efforts to control for the size of buildings and their use hopefully keep many of these factors under control if they do in fact vary with climate.

12.2.4   Estimation

The probability of cooling is estimated with a logit regression (equation 12.2). Insignificant coefficients are dropped from the model. Short-run and long-run models are then estimated to measure how energy expenditures in residential and commercial buildings vary with climate, demographic and firm-specific information, and building characteristics. The predicted value of cooling is also entered into this second-stage equation.

Regressing building characteristics on climate determines which building characteristics are climate-sensitive. The results show that the size of the home, the number of doors and windows, a wood-burning stove, a basement, inadequate insulation, discounted electricity rates, central air-conditioning, air-conditioning units, electric radiator heat, and the availability of natural gas are all residential characteristics that vary with climate. The empirical work also identifies that air ducts for heating, air ducts for cooling, boilers, heat pumps for cooling, build-up of roof material, metal roofs, shingle walls, the number of floors, and the use of alternative fuels are all commercial characteristics that vary with climate. Note that some of these characteristics reflect overall space conditioning energy demand, some affect insulation, and some reflect heating and cooling capacity.

A log-log functional form is then used to estimate energy expenditures. This functional form: (1) provides the highest predictive power based on F-tests of the overall significance of the regression; (2) is commonly used in the energy demand literature; and (3) yields the expected proportional relationship between the continuous variables and energy expenditures. The climate variables are included in both linear and squared form to reflect the hypothesized quadratic climate-expenditure relationship. The mean of the climate variables has been subtracted from each climate measure. Consequently, the estimated coefficient (β_0) on the linear term (C) in the regression model (equation 12.7) can be interpreted as the marginal impact of that climate variable on energy expenditures evaluated at the mean of the sample.

The hypothesized expenditure equation for the short-run demand for energy is:

$$\ln \sum_{i=1}^{F} E_i = \alpha + \beta_0 C + \beta_1 C^2 + \beta_2 \ln P + \beta_3 \ln S + \beta_4 \ln Z_{nc}$$

$$+ \beta_5 D_{nc} + \beta_6 \ln Z_c + \beta_7 D_c \qquad (12.8)$$

where $i = 1$, F represents the total number of fuels, E_i is energy expenditures, C is a vector of climate variables, P is a vector of average fuel prices, S is a vector of demographic (firm-specific) characteristics, Z is a vector of building characteristics, and D is a vector of dummy variables. Subscripts c and nc represent portions of the vectors Z and D that are climate sensitive and non-climate sensitive, respectively.

We explored several alternative methods of measuring climate in this analysis. Following the literature, we first examined using cooling and heating degree-days. Unfortunately, heating (cooling) degree-days are based on a deviation from an arbitrary base temperature. The degree-days automatically go to zero when the temperature goes to the base level. By assumption, these constructs force the minimum expenditure point to the assumed base temperature. Further, we were eager to relate our energy results to the outcomes of the climate models. We consequently turned to using monthly normal temperature and precipitation to measure climate. These are 30 year averages of temperature and precipitation for a specific month. We explored using January, April, July, and October, but found the four monthly variables to be highly correlated. We followed Morrison and Mendelsohn (1999) for the commercial sector and used the annual temperature and the difference between winter and summer. For the residential sector, we used winter and summer climate, the temperature and precipitation in January and July. We found that these two different climate measures give the most sensible results for these two sectors.

The long-run model omits the climate-sensitive building characteristics, allowing these characteristics to be endogenous. The resulting impact captures how energy expenditures would change as building characteristics change freely as well:

$$\ln \sum_{i=1}^{F} E_i = \alpha + \beta_0 C + \beta_1 C^2 + \beta_2 \ln P + \beta_3 \ln S + \beta_4 \ln Z_{nc} + \beta_5 D_{nc} \qquad (12.9)$$

These equations are estimated using microdata for individuals and firms. Because the data come from individual households and firms, there should be no problems with identification. This contrasts with many energy studies that rely on aggregate data, where prices would be endogenous. With disaggregated data, we can reasonably assume that prices are exogenous to each home or firm and that prices are given. However, prices can be a problem in locations with non-linear price schedules. For this study, we used average prices calculated by dividing energy expenditures by quantities consumed. This may lead to some error in the calculated price elasticities. However, it should not bias the climate coefficients.

12.3   RESULTS

12.3.1   Residential Results

Table 12.1 presents the logit results for the probability that central cooling is installed in residential properties. The coefficients in Table 12.1 are generally well behaved. Higher prices for fuels used for cooling reduce the probability of cooling. Newer buildings are much more likely to cool. Larger buildings are more likely cooled. Home-owner attributes such as higher income and age (over 65) increase the probability of cooling, but factors such as the tenant controlling the heat, the age of the head of household, cash aid, and family size all decrease the probability of cooling. Residential building characteristics such as the number of units, the number of floors, color TV, and more appliances all increase the probability the building is cooled.

Table 12.1 suggests that, as expected, residential cooling increases sharply with summer temperatures. The probability of cooling, however, increases at a decreasing rate as temperatures rise. At mean winter temperatures in the United States, the marginal effect of temperature of cooling is not different from zero. However, the significant squared coefficient implies that the probability of cooling increases at an increasing rate as winter temperatures warm

Table 12.1   Logit regression of residential probability of cooling

Variable	Coefficient (t-statistic)	Variable	Coefficient (t-statistic)
Constant	797 (20.41)	Log # floors	0.21 (1.92)
January temperature	−9.03e−3 (1.27)	Log family size	−0.78 (9.34)
January temperature^2	4.50e−3 (6.46)	Log age of household head	−0.31 (2.15)
July temperature	0.43 (20.41)	Multiple units	0.87 (6.34)
July temperature^2	−3.79e−2 (9.36)	Tenant controls heat	−0.95 (2.27)
Log electricity price	−0.76 (4.88)	Over 65	0.26 (1.92)
Log natural gas price	−0.84 (5.47)	Cash aid	−1.31 (2.11)
Log year built	103 (20.11)	TV color	0.36 (7.98)
Log income	0.63 (10.52)	Appliances	0.88 (6.42)
		Log of rooms	1.81 (11.07)
Observations	5030	Log likelihood	−2166

above the mean. Winter temperatures have a role in cooling because they reflect how long temperatures will remain warm throughout the year. The length of the hot season figures prominently in Americans' choice of cooling equipment. Precipitation has no effect on the probability of residential cooling.

Table 12.2 presents the regression results for the short-run and long-run versions of the residential model. The dependent variable is the log of total energy expenditures. The control variables in the residential sector are well behaved. Because the dependent variable is expenditures, not quantity demanded, price elasticity must be calculated from the estimated coefficients. Note that predicted elasticities based on average rates are expected to be biased, but Halvorsen (1975) demonstrated that with a double-log form, elasticity estimates for marginal and average rates are quite comparable. Two-stage least-squares regression analysis was used to test for the importance of non-linear price schedules. Prices were regressed on a set of demand variables and then the predicted price was used in the

Table 12.2   Regression model of residential energy expenditures

Variable	Short run	Long run	Variable	Short run	Long run
Constant	42.7	40.2	Log income	5.39e−2	7.88e−2
	(8.27)	(8.65)		(7.64)	(11.05)
January	−7.60e−3	−9.37e−3	Log # floors	−9.8e−2	−0.12
temperature	(4.63)	(5.60)		(7.68)	(9.23)
January	−2.23e−4	−0.97e−4	Log family size	0.23	0.22
temperature^2	(1.90)	(0.80)		(23.64)	(22.17)
January	1.87e−2	2.30e−2	Log age of head	5.47e−2	7.69e−2
precipitation	(3.57)	(4.24)		(3.93)	(5.40)
January	−1.63e−3	−1.84e−3	Multiple units	−9.70e−2	−9.95e−2
precipitation^2	(2.39)	(2.62)		(5.81)	(6.32)
July temperature	1.19e−2	1.87e−2	Tenant controls	0.17	0.22
	(3.49)	(5.47)	heat	(4.31)	(5.17)
July temperature^2	2.48e−3	2.43e−3	Hispanic	−4.61e−2	−6.20e−2
	(4.96)	(4.74)		(2.46)	(3.19)
July precipitation	2.03e−2	2.94e−2	Cash aid	−0.19	−0.20
	(4.99)	(7.09)		(3.90)	(3.85)
July precipitation^2	−4.56e−3	−5.95e−3	Heat aid	6.69e−2	6.03e−2
	(2.95)	(3.72)		(2.59)	(2.25)
Log electricity	0.33	0.33	TV color	6.09e−2	7.32e−2
price	(17.88)	(17.57)		(11.09)	(12.99)
Log natural	0.17	0.19	Computer	4.00e−2	5.14e−2
gas price	(8.58)	(9.53)		(3.03)	(3.75)
Log LPG price	−8.9e−2	−7.9e−2	Appliances	0.10	0.14
	(4.02)	(3.45)		(7.12)	(9.64)
Log year built	−4.96	−4.55	Log of rooms	0.12	0.36
	(7.31)	(7.41)		(5.02)	(18.94)
Log home area	0.018	−	Electricity	−7.17e−2	−
	(12.38)		discounts	(2.61)	
Wood burning	−5.09e−2	−	Central air	0.15	−
stove	(4.33)		conditioning	(11.03)	
Log doors	8.41e−2	−	Wall or window	4.10e−2	−
and windows	(6.02)		air conditioner	(3.50)	
Basement	−9.47e−2	−	Electric wall	9.05e−2	−
	(7.14)		heater	(6.80)	
Poor insulation	4.04e−2	−	Natural gas	−4.96e−2	−
	(3.42)		available	(4.40)	
Observations	5030	5030	Adjusted R^2	0.97	0.96

Note: T-statistics are in parentheses.

expenditure regression. The 2SLS models had similar climate coefficients to the model shown but the price coefficients behaved poorly.

The estimated price elasticities in Table 12.2 fall within the predicted range of the literature from -0.5 to -2.0 (Wilson, 1971; Anderson, 1973; Halvorsen, 1975; Barnes et al., 1981; Baker et al., 1989). The residential estimates are -0.7 for electricity, -0.8 for natural gas, and -1.1 for liquefied petroleum gas (LPG).

Demographic characteristics that positively influence residential energy expenditures include income, family size, age of household head, tenant controlling the heat, and receipt of heating vouchers. Note that having the tenant control the heat is not the same as having the tenant pay for heat separately. Expenditures are less if the head is Hispanic, receives cash aid for heating (measure of poverty), participates in an energy discount program, or burns wood as an alternative fuel. Structural characteristics that positively influence expenditures include home area, the number of rooms, the number of doors and windows, the age of the house, and inadequate insulation. The presence of a basement, the more residential units in the property, and the more floors all reduce energy expenditures. Appliances and electrical equipment such as a TV, a computer, a dishwasher, a clothes washer, and a dryer increase energy expenditures. Space conditioning equipment, including central air-conditioning, wall or window air-conditioners, and electric wall or radiator heaters, increase expenditures. Only some households have access to natural gas, but if this option is available, household expenditures are less.

The following variables were found to be climate sensitive in the residential analysis: home area, wood-burning stove, number of doors and windows, basement, poor insulation, discount electricity, central air-conditioning, window air-conditioning, electric wall heat, and natural gas availability. These variables were included in the short-run regression but omitted from the long-run regression. The long-run model allows these variables to change endogenously with climate.

Residential climate is represented in terms of January and July temperature (°C) and precipitation (inches per month). Alternative specifications yielded inferior results. For example, with annual temperature and precipitation and winter–summer differences, precipitation was not significant if there were no clear winter and summer effects (Morrison and Mendelsohn, 1999). Adding April and October climate variables did not produce sensible coefficients. Using four correlated seasons is probably too complicated for an energy model.

In the residential sector (Table 12.2), warmer January temperatures reduce expenditures on the margin in both the short- and long-run models. Warmer winter temperatures are expected to reduce expenditures for heating. July temperatures have a U-shaped relationship with energy expenditures with a cost-minimizing temperature of 20°C. California is a few degrees warmer than this minimum temperature. Warming in the summer, then, will increase energy expenditures throughout most of California. Because the summer increase is larger than the winter decrease, warming will cause a net increase in annual energy expenditure for the average California residence. Both January and July precipitation positively influences expenditures at the mean, suggesting that more humid locations have higher energy expenditures. The effect of higher humidity increases the demand for both heating in the winter and cooling in the summer, and it also increases the cost of both. The squared terms for summer and winter precipitation are significant and negative, suggesting that this effect diminishes as humidity increases.

12.3.2 Commercial Results

Table 12.3 presents the results of the logit cooling regression for the commercial sector. Higher prices for both electricity and natural gas reduce the probability of cooling because these are the two fuels used for cooling. More modern buildings, as well as buildings used for food, health, laboratories,

Table 12.3 Logit regression of commercial probability of cooling

Variable	Coefficient (t-statistic)	Variable	Coefficient (t-statistic)
Constant	−139 (7.30)	Log year built	17.77 (7.04)
January temperature	−4.29e−2 (4.38)	Food sales and service	9.76e−3 (6.59)
January precipitation	0.14 (6.57)	Warehouse	−6.98e−3 (7.17)
July temperature	0.22 (12.66)	Health facility	11.6e−3 (5.34)
July precipitation	−0.13 (6.92)	Outpatient health	9.51e−3 (4.46)
July precipitation^2	4.14e−2 (4.18)	Laboratory	8.68e−3 (2.92)
Log electricity price	−0.25 (2.88)	Office	12.5e−3 (13.51)
Log natural gas price	−0.31 (4.17)	Retail	28.6e−3 (3.31)
Log square feet	0.33 (16.44)	Urban area	0.15 (1.94)
Observations	5611	Log likelihood	−3231

offices, and retail, are more likely to cool. Interestingly, larger buildings are more likely to be cooled despite the fact that they are more expensive to cool. This implies that many buildings that are not cooled house relatively small operations such as neighborhood offices and retail spaces. Buildings such as warehouses are also often not cooled.

Table 12.3 also reveals the connection between commercial cooling and climate. Higher July temperatures increase the probability of cooling, but higher January temperatures actually reduce cooling probabilities. The commercial sector, relative to the residential sector, seems more responsive to the severity of temperature in the summer rather than the length of the period it is hot. Cooling in the commercial sector is also sensitive to precipitation. Higher summer precipitation has a U-shaped relationship with cooling, at first reducing cooling, but then increasing it. Higher winter precipitation increases cooling. These relationships may reflect the interaction between heat and humidity. The higher the humidity, the more unpleasant is high heat, and thus the greater the demand for cooling.

Table 12.4 presents the short-run and long-run commercial results of the expenditure analysis. The commercial price elasticity estimates fall within the range most commonly cited in the literature, -1.0 to -3.0 (Baughman and Joskow, 1976; Mount et al., 1993). The specific price elasticities predicted in this study are -1.5 for electricity and -1.1 for natural gas. Characteristics of the commercial operation that positively influence expenditures include the number of months the business is open per year, the various building uses, and the presence of appliances such as ice, water, or vending machines and commercial refrigerators and freezers. Warehouses that are not refrigerated and tenants that control the heat, for example, reduce energy expenditures. Building characteristics that increase energy expenditures include square footage, number of floors, and built-up roof material. Buildings that are more modern use more energy, which is an interesting result because it suggests that the greater demand for energy from new equipment exceeds any efforts to install more insulation and other conservation features. Glass and metal roofing materials lower expenditures, as do wall materials such as masonry and shingles. All space conditioning equipment causes expenditures to be higher, except heat pumps for cooling, which lower expenditures. Buildings that use alternative fuels that are not included in the calculation of energy costs have lower energy expenditures.

In the commercial analysis, the following variables are climate sensitive: air ducts for cooling, air ducts for heating, boilers, heat pumps for cooling, built-up roof material, metal roofs, number of floors, and alternative fuel use. These variables are included in the short-run regression but are excluded from the long-run regression.

Table 12.4   Regression model of commercial energy expenditures

Variable	Short run	Long run	Variable	Short run	Long run
Constant	−50.2	−35.9	% laboratory	9.89e−3	9.88e−3
	(5.63)	(5.21)		(6.71)	(6.48)
Annual	−1.67-2	−1.69e−2	% industry	13.4e−3	13.7e−3
temperature	(3.26)	(3.39)		(3.96)	(3.91)
Annual	3.28e−3	3.59e−3	% office	4.22e−3	5.68e−3
temperature^2	(4.60)	(4.90)		(9.62)	(12.81)
Summer–winter	1.33e−2	1.59e−2	% retail	2.81e−3	2.74e−3
temperature	(2.90)	(3.37)		(7.51)	(7.33)
Summer–winter	−0.62e−4	−0.56e−4	% education	2.76e−3	3.44e−3
temperature^2	(0.17)	(0.15)		(4.89)	(5.98)
Log electricity	−0.50	−0.50	Metropolitan	0.35	0.45
price	(16.05)	(15.65)		(12.47)	(16.16)
Log natural	−0.08	−0.12	Ice machine	0.49	0.54
gas price	(2.78)	(3.97)		(17.73)	(18.87)
Masonry wall	−0.76	−0.73	Commercial	0.45	0.49
	(4.05)	(3.74)	refrigerator	(12.54)	(13.30)
Log square foot	0.53	0.59	Computer	0.58	0.64
	(39.63)	(46.33)	cooling	(10.90)	(11.63)
Months open/year	2.75e−2	2.49e−2	Heat pump	0.14	8.63e−2
	(4.99)	(4.38)	for heat	(2.08)	(2.10)
Log year built	6.71	4.76	Tenant controls	−0.14	−0.16
	(5.69)	(5.23)	heat	(5.21)	(6.08)
% food sale/serve	8.31e−3	8.54e−3	Roof glass	−1.77	−1.95
	(13.67)	(13.69)		(5.61)	(5.98)
% nonrefrigerated	−3.27e−3	−4.01e−3	California	0.37	0.36
warehouse	(7.23)	(8.89)	dummy	(5.58)	(5.24)
% health	7.04e−3	8.33e−3	% outpatient	2.96e−3	3.82e−3
	(5.81)	(6.67)		(3.61)	(4.52)
Air duct cooling	−22.8	–	Roof-metal	−0.15	–
	(1.69)		surface	(4.37)	
Air duct heating	6.97e−2	–	Wall shingle	−0.17	–
	(2.12)			(5.03)	
Boilers	0.24	–	Log floors	0.14	–
	(6.64)			(4.30)	
Heat pump cooling	−0.14	–	Alternative fuel	−0.54	–
	(1.94)			(11.89)	
Roof built-up	0.13	–	Cooling*age	3.02	–
	(4.78)			(1.70)	
N	5611	5611	Adjusted R^2	0.96	0.96

Note: T-statistics are in parentheses.

According to Table 12.4, warmer annual temperatures decrease energy expenditures at the US mean. The squared term on temperature, however, is positive so that energy expenditures have an overall U-shape with respect to temperature. The cost-minimizing annual temperature is 15.8°C, which is 2.4°C above the US average but below the average temperature for California. On average, warming will increase California commercial expenditures. The difference between summer and winter temperatures increases energy expenditures as expected. The greater this difference, the more that has to be spent on heating in the winter and cooling in the summer.

The cost-minimizing temperature for the commercial sector is warmer than the cost-minimizing temperature for the residential sector, and it is higher than what engineering models would suggest. Engineering studies suggest that commercial buildings have substantial waste heat from lights and other activities. The engineering studies predict that the minimum energy expenditures for commercial buildings should be less than those for residences. The cross-sectional observations, however, reflect the fact that residential owners do not invest in air-conditioning in the cooler north. Consequently, the cooler north has less residential energy expenditures than the south. Commercial buildings in the north, in contrast, do install air-conditioning, so the minimum energy location in the commercial sector is actually much farther south (in warmer locations).

We examined a few other analyses of climate–energy interactions to test the robustness of the empirical results. For example, we limited the data to firms and households in the Pacific region and re-estimated the equations. Although the results were quite similar to the national results, many of the coefficients were less significant because of the reduced sample size. We also estimated the equations using data that are likely to have come from California, but this led to unsatisfactory results for two reasons – we cannot determine precisely which state each observation comes from, and California has a limited range of climate variation. Consequently, the analysis relies on the national empirical results.

12.3.3 Climate Change Simulations

In the following simulations, a baseline is first developed to estimate what energy expenditures might look like in the future based only on socio-economic changes. We examine both the slow and fast economic paths from Chapter 2. As California's population expands, we assume that energy consumption expands proportionately. We expand the population in each county using a high and low projection of state population in 2020, 2060, and 2100 as discussed in the baseline (Chapter 2). We also make an adjustment for changes in buildings. As new buildings replace older ones, we

Table 12.5   Projected baseline energy expenditures for California ($ billion)

Sector	Year		
	2020	2060	2100
Residential			
Slow growth	20.9	30.9	36.7
Fast growth	21.5	38.4	61.2
Commercial			
Slow growth	15.1	16.9	15.7
Fast growth	15.1	19.7	24.0

Note: Scenarios include population growth, real energy price increases, income per capita, updated buildings, and increased cooling. Baseline assumes current climate.

expect that new energy features will replace old technologies. We capture this effect by updating the time that buildings are built in each time period. We expect that energy prices will increase over time as inexpensive deposits of fossil fuels become rare. We build a 1 percent per year increase into all fuel prices; for example, Manne and Richels (1992) assume a 1.7 percent price increase in their business-as-usual case for non-electric energy, although they assume electric prices will remain constant. Because future energy prices are difficult to predict, we also include a sensitivity analysis with no real price increases.

Economic development is expected to have a strong effect on baseline energy projections for the state without climate change. The baseline energy expenditures for each sector are projected in Table 12.5. The expenditures by 2020 are somewhat similar since not much time has yet passed. However, by 2100, expenditures on energy exhibit quite a large range because of the difference between the slow and fast economic growth rates.

We expect that cooling will continue to penetrate both the residential and commercial markets. Using the logistic functions in Tables 12.1 and 12.2, we then predict the level of cooling that will result from the population, income, modern buildings, and prices in future periods. Cooling is predicted to penetrate more completely in the residential market, reaching more than 80 percent of homes in California by the end of the century (see Table 12.6). In contrast, only a little more than half of the commercial buildings are expected to install cooling by 2100. The fact that cooling does not penetrate more completely into the commercial sector reflects the wide set of uses for commercial buildings. For example, warehouses are not likely to be air-conditioned at all. Changes in cooling behavior are captured in the analysis by weighting future samples toward buildings with cooling.

Table 12.6   Percentage of future buildings with cooling in California

	Residential	Commercial
2020		
Slow growth	43.7	47.0
Fast growth	46.2	47.0
2060		
Slow growth	65.7	49.6
Fast growth	70.9	49.6
2100		
Slow growth	82.8	52.2
Fast growth	87.9	52.2

Source: From Tables 12.1 and 12.2.

Relying on the estimates from Tables 12.1 and 12.3, warming is expected to increase cooling in both residential and commercial buildings. With the three uniform incremental warming scenarios for 2100, the percentage of buildings that are cooled in the state clearly increases the warmer the scenario. The residential sector is particularly responsive, increasing cooling installation in up to 99 percent of homes in the most severe warming scenario. Even the commercial sector is responsive, increasing cooling installation in up to 70 percent of the state's buildings in the most severe warming case.

For each climate prediction, the estimated regression equations in Tables 12.1 and 12.2 are used to predict how residential energy expenditures will change and the equations from Tables 12.3 and 12.4 are used to predict how commercial energy expenditures will change. The new energy expenditures from each climate scenario are compared to the baseline energy expenditures for the same year. The change in energy expenditures is a measure of the welfare effect of the climate change. Increases in energy expenditures are losses and decreases in energy expenditures are benefits.

With the uniform climate scenarios, the impacts of warming are expected to cause damages in the residential sector in every warming scenario. These damages increase the warmer the scenario. For example, with slow growth, no precipitation increase, and the 1.5°C warming, long-run damages are $1.5 billion per year, but with the same conditions and a 5°C warming, long-run damages rise to $6 billion per year. The damages increase slightly when precipitation increases but the damages are quite sensitive to the baseline assumptions about economic growth. For example, the 5°C warming case leads to $6 billion with slow growth but $9 billion with faster growth.

Comparing the short-term and long-term results suggests that long-term impacts will generally be larger than those seen in the short term.

Table 12.7   *Annual California welfare impacts from general circulation model scenarios ($ million)*

		Residential		Commercial	
		Short run	Long run	Short run	Long run
Hadley					
2020	Slow	−1824	−1589	−85	−140
	Fast	−1883	−1653	−85	−140
2060	Slow	−3589	−3369	−644	−864
	Fast	−4091	−3954	−840	−1127
2100	Slow	−4219	−4740	−1973	−2522
	Fast	−5915	−7099	−3553	−4543
PCM					
2020	Slow	−296	−189	−47	−64
	Fast	−310	−200	−47	−64
2060	Slow	−2547	−2234	−453	−611
	Fast	−2903	−2615	−590	−797
2100	Slow	−3067	−2947	−1172	−1534
	Fast	−4010	−4126	−2111	−2763

Note: The negative numbers are damages (increases in energy. expenditures).

When buildings can adjust to the warmer climates, the economic damages increase. This predicted increase in expenditures in the long run is caused by the increase in cooling capacity in the long run. The difference in energy expenditures between the long run and short run, however, is not that large. It would appear that building adjustments are not that important in the residential model.

The commercial results using the uniform scenarios are quite similar to the residential results. For mild warming, there are damages ranging from $300 to $900 million. With a large warming such as 5.0°C, damages rise to $3.9 to $8.7 billion. However, energy expenditures are not similar in the short-run and long-run commercial models. Long-run damages are much larger than short-run damages. Climate-sensitive building adjustments are predicted to result in much higher energy expenditures because of the increase in central cooling.

The residential impacts of the GCM scenarios are presented in Table 12.7. Each GCM model provides estimates of climate change in 2020, 2060, and 2100. The resulting impacts fall within the range of impacts predicted by the uniform scenarios. The very wet Hadley scenario results in small damages in 2020 because climate has changed only slightly by then.

However, by 2100, the impacts predicted by the Hadley model fall between the 'middle-of-the-road' wet and the more severely wet incremental scenarios. The PCM model predicts much gentler impacts both because precipitation declines and temperatures do not increase rapidly. By 2100, the impacts from the PCM fall between the middle of the road and the mildly dry incremental scenarios.

Table 12.7 also presents the commercial results for the GCM scenarios. The Hadley and PCM models predict relatively small damages in 2020. These annual damages rise to $864–$1127 million in the Hadley model by 2060 and to $611–$797 million in the PCM model as temperatures rise. The faster the energy sector grows, the greater are the damages from warming, especially over time.

A sensitivity analysis was also conducted which tests the importance of projecting growth into the future and the price assumptions. If the energy sector does not grow in the future, the impacts fall to about one-half of the projected impacts in the slow-growth scenario. If the economy grows but the prices of energy remain constant, the damages from warming also fall dramatically. This effect is especially strong in the residential sector because it is price inelastic. The lower prices cause residential expenditures in energy to shrink, reducing the damages from climate change.

12.4   CONCLUSIONS

This study estimates a two-equation model of energy using a cross-section of American households and firms. The method section explains how this model can be used to measure the welfare impacts of a change in climate. If we assume that people maintain a desired interior temperature, then any predicted change in energy expenditure would measure the welfare effect of climate change. Increases in energy expenditures would be damages and reductions in expenditures would be benefits.

The empirical model has both a cooling and an expenditure equation. A separate regression was estimated for residential and commercial properties. This study improves on past cross-sectional studies by explicitly modeling the probability of cooling. We feel that this improvement mitigates one of the biases of the earlier studies, which did not take changes in summer interior temperatures into account.

The observed empirical relationship between climate and energy was expected. As winter temperatures rise, residential energy expenditures on heating fall. As summer temperatures rise, residential energy expenditures on cooling increase. Similarly, as average annual temperatures increase, commercial energy expenditures at first fall, but then rise as well, following a U shape. Theory predicts this observed U-shaped relationship between annual energy expenditures and outside temperature.

This work tests the importance of changes in buildings as part of the climate sensitivity of energy expenditures. We estimated both a short-run and a long-run model of energy expenditures. The short-run model freezes building characteristics in place. The long-run model allows climate-sensitive building characteristics to change with climate. By comparing the results of the two models, we can test whether the building changes are important. In both the residential and commercial sectors, the long-run damages are somewhat larger than the short-run damages. In the long run, both sectors spend more on energy as cooling capacity is increased. Because the long-run expenditures are greater than the short-run costs, they are probably the more accurate estimate of the actual damages. The difference between short- and long-run results was small for the residential sector but larger for the commercial sector. The results suggest that it is important to model how commercial buildings might change as the climate warms.

We also tested how climate affects the probability that residential and commercial buildings have central cooling. Using logit analysis, we find that buildings are more likely to be cooled if they are in warmer climates. These empirical estimates are included in our analysis by weighting future samples of buildings to include more cooled structures.

Combining information about energy expenditures with the information about future conditions, we were able to predict how warming might affect California in the future. Warming is predicted to increase energy expenditures in residential and commercial buildings. With mild warming (1.5°C), the state damages range between $1.3 and $3.5 billion per year by 2100. With warming of 3°C, the damages range between $3.5 and $8.7 billion annually. Finally, with more warming (5°C), the damages range between $7.7 and $17.8 billion annually. Two factors explain why the range of impacts is so large: the future growth of the state's economy and the future prices of energy.

The residential energy impacts across counties in the state are not uniform. Plate 9 maps the change in residential energy expenditures for the 3°C uniform scenario with an 18 percent increase in precipitation in 2100. Reductions in energy expenditures are benefits and increases are damages. The impacts vary depending on the initial climate of the county. For example, the cool northern maritime and high alpine counties exhibit small decreases in energy expenditures (benefits) from warming. The hot southern desert counties have the largest increases in energy expenditures (damages). The remaining central valley and southern maritime counties have average state energy increases.

Plate 10 maps the change in commercial energy expenditures from the same 3°C, +18 percent precipitation uniform scenario. The results in the commercial sector are similar to the residential results. The northern maritime and alpine counties benefit from warming. The southern desert counties exhibit the largest damages per building. The central valley also suffers damages from warming, although these effects are smaller than in the desert. The southern maritime region also has small damages from warming.

This analysis gives a sense of the importance of climate change to the energy sector. The analysis suggests that California will experience an increase in both commercial and residential energy expenditures from warming. The resulting damages will increase as the warming becomes more severe. The size of the net impacts also depends on baseline assumptions about growth and energy prices. If warming occurred today or there was no growth in the size of the energy sector, damages would be about one-third of what is forecast for 2100. The impacts vary across counties, with the southern desert counties suffering the largest damages and the Central Valley also being heavily damaged. The northern maritime and alpine counties may well have small benefits from warming.

REFERENCES

Anderson, K.P. 1973. Residential Energy Use: An Econometric Analysis. R-719-NF. The Rand Corporation, Santa Monica, CA.

Baker, P., R. Blundell, and J. Micklewright. 1989. Modeling household energy expenditures using micro-data. *Economic Journal* **99**: 720–38.

Barnes, R., R. Gillingham, and R. Hagemann. 1981. The short run residential demand for electricity. *Review of Economics and Statistics* **63**: 541–52.

Baughman, M. and P. Joskow. 1976. Energy consumption and fuel choice by residential and commercial consumers in the United States. *Energy Systems and Policy* **1**(4): 305–23.

Baxter, L. and K. Calandri. 1992. Global warming and electricity demand. *Energy Policy* 233–44.

Belzer, D., M. Scott, and R. Sands. 1996. Climate change impacts on US commercial building energy consumption: An analysis using sample survey data. *Energy Sources* **18**: 177–201.

Cline, W. 1992. *The Economics of Global Warming*. Institute for International Economics, Washington, DC.

Crocker, T. 1976. Electricity demand in all-electric commercial buildings: The effect of climate. In *The Urban Costs of Climate Modification*, T. Ferrar (ed.). John Wiley & Sons, New York, pp. 139–56.

Dai, A., G.A. Meehl, W.M. Washington, T.M.L. Wigley, and J.M. Arblaster. 2001b. Ensemble simulation of 21st century climate changes: Business as usual vs. CO_2 stabilization. *Bulletin of the American Meteorological Society* **82**: 2377–88.

Dai, A., T.M.L. Wigley, B.A. Boville, J.T. Kiehl, and L.E. Buha. 2001a. Climates of the twentieth and twenty-first centuries simulated by the NCAR Climate System Model. *Journal of Climate* **14**: 485–519.

Energy Information Administration. 1992. *Commercial Buildings Energy Consumption and Expenditures*. DOE/EIA-0318. US Department of Energy, Washington, DC.

Energy Information Administration. 1993. *Household Energy Consumption and Expenditures*. DOE/EIA-0321. US Department of Energy, Washington, DC.

Halvorsen, R. 1975. Residential demand for electric energy. *Review of Economics and Statistics* **57**: 13–18.

Johns, T.C., R.E. Carnell, J.F. Crossley, J.M. Gregory, J.F.B. Mitchell, C.A. Senior, S.F.B. Tett, and R.A. Wood. 1997. The second Hadley Centre coupled ocean-atmospheric GCM: Model description, spinup and validation. *Climate Dynamics* **13**: 103–34.

Linder, K.P., M.J. Gibbs, and M.R. Inglis. 1989. *Potential Impacts of Climate Change on Electric Utilities*. EPRI EN-6249. Electric Power Research Institute, Palo Alto, CA.

Manne, A. and R. Richels. 1992. *Buying Greenhouse Insurance: The Economic Costs of CO_2 Emission Limits*. MIT Press, Cambridge, MA.

Mendelsohn, R. 2001. Energy: Cross-sectional analysis. In *Global Warming and the American Economy: A Regional Assessment of Climate Change Impacts*, R. Mendelsohn (ed.). Edward Elgar, Cheltenham, UK and Northampton, MA, pp. 149–66.

Mendelsohn, R. and J. Neumann. 1999. Synthesis and conclusions. In *The Impact of Climate Change on the United States Economy*, R. Mendelsohn and J. Neumann (eds), pp. 315–31. Cambridge University Press, Cambridge, UK.

Mendelsohn, R., W. Nordhaus, and D. Shaw. 1994. The impact of global warming on agriculture: A Ricardian analysis. *American Economic Review* **84**(4): 753–71.

Metz, B., O. Davidson, R. Swart, and J. Pan. 2001. *Climate Change 2001: Mitigation*. Third Assessment Report of the Intergovernmental Panel on Climate Change. Cambridge University Press, Cambridge, UK.

Morrison, W. and R. Mendelsohn. 1999. The impact of global warming on US energy expenditures. In *The Impact of Climate Change on the United States Economy*, R. Mendelsohn and J. Neumann (eds). Cambridge University Press, Cambridge, UK, pp. 209–36.

Mount, T.D., L.D. Chapman, and T.J. Tyrrell. 1993. Electricity Demand in the United States: An Econometric Analysis. ORNL-NF-49. Oak Ridge National Laboratory, Oak Ridge, TN.

Nelson, J. 1976. Climate and energy demand: Fossil fuels. In *The Urban Costs of Climate Modification*, T. Ferrar (ed.). John Wiley & Sons, New York, pp. 123–37.

Nordhaus, W. 1991. To slow or not to slow: The economics of the greenhouse effect. *Economic Journal* **101**: 920–37.

Rosenthal, D., H. Gruenspecht, and E. Moran. 1995. Effects of global warming on energy use for space heating and cooling in the United States. *Energy Journal* **16**(2): 77–96.

Sailor, D. 2001. Relating residential and commercial sector electricity loads to climate. *Energy* **26**: 645–57.

Smith, J. and D. Tirpak. 1989. *The Potential Effects of Global Climate Change on the United States*. US Environmental Protection Agency, Washington, DC.

Wilson, J.W. 1971. Residential demand for electricity. *Quarterly Review of Economics and Business* **11**(1): 7–22.

13.   Coastal impacts

James Neumann and Daniel Hudgens

13.1   INTRODUCTION

The potential effects of climate change on sea level are well established (IPCC, 2001); a global temperature increase is likely to lead to both thermal expansion and the melting of polar ice caps, which contribute to sea level rise. Increases in sea level can affect individuals living in coastal and low-lying areas, inundate or impair wetland functions, and damage structures and property along the coast. This chapter examines how sea level rise affects the developed coastline of California and how California can adapt to this problem over time.

Research on the economic costs of sea level rise began in the 1980s, and progressed from early assessments that focused on total property at risk and protection costs (for example Gleick and Maurer, 1990; Titus et al., 1991) in isolation to more recent efforts that evaluate both potential lost property values and protection costs over time (Yohe et al., 1996, 1999). Many of the published estimates focus on a very limited geographic scope or are national in scope. It has recently become clear that regional and sub-regional estimates of the economic cost of sea level rise are also needed to inform coordinated response and adaptation planning that is responsive to local conditions but recognizes that other relevant coastal policy choices may be made at the state level.

The California Coastal Commission (2001) examined the potential impacts of sea level rise along the state's coast. The commission identified significant portions of coast that were both extensively developed and low-lying or prone to erosion, including San Diego; Los Angeles; portions of San Luis Obispo, Monterey, and San Mateo; and Humboldt County. The commission also concluded that a mix of hard engineering (for example armoring) and soft engineering (for example beach nourishment) is likely to be employed, but retreat from certain threatened areas is also an option. The study, however, did not estimate the economic cost of these responses.

In this chapter, we assess the economic costs of sea level rise statewide for California. Our research focuses on effects on coastal structures; we do not examine the impacts on wetlands and some undeveloped dryland areas because of a lack of data and methods to evaluate these resources. The value of such lands is not included in the estimates of the costs of sea level rise. Also, we did not estimate the costs of protecting the Sacramento Delta islands from sea level rise. These islands are already protected by levees, and either the levees would have to be raised or entire islands would be allowed to be submerged.

To estimate the cost of sea level rise for California, we must consider three basic types of information: (1) when and where will areas be inundated; (2) how will property owners and governments respond; and (3) what are the costs of plausible alternative responses. Our work relies on inundation mapping at sites modeled by Park et al. (1989) to estimate both the location and timing of inundation for four sea level rise scenarios: 33, 50, 67, and 100 cm by 2100.[1] The second of these (50 cm) closely approximates the expected rate of eustatic sea level rise established in IPCC (2001). Using the approach developed in Yohe et al. (1999), we model both the response and the cost of that response.

The economic model we use has three significant advantages over previous work. First, it incorporates a site-specific decision-making process to assess whether it is more efficient to protect or abandon specific parcels of land, based on the costs of protection and the value of coastal structures that could be protected. Second, it incorporates changes in property value over time, based on a representation of property value that relies on projections of gross domestic product (GDP) and population growth. Third, it incorporates adaptive measures that landowners could take to mitigate impacts in the coastal zone, including ceasing investment in coastal properties in advance of inundation. Implicit in our characterization of adaptive measures is that adaptation is efficient; that is, we assume that individuals and the government protect only when the benefits of protection exceed the costs and that they act at the optimal moment. This is an optimistic assumption in this sector because the most likely protective measures are 'public adaptations', where there are many beneficiaries to an action. In practice, public adaptation that requires collective action is not likely to be efficient (Mendelsohn, 2001).

It is also important to note that our estimates reflect a focus on the effects of inundation on coastal property and the costs of protection. We did not examine the protection of wetlands or the incremental storm damage caused by changing storm intensities or frequencies or by higher seas.

In the next section, we review our method for developing the estimates, including the selection and coding of data for new sites that, in aggregate, span a diverse set of conditions along the California coast. The third section discusses the site-specific and state-level results, including transient estimates of the timing of impacts for each region. Finally, we present the conclusions, key uncertainties inherent in the approach, and steps that might be taken to improve the estimates.

13.2 METHOD

The Yohe et al. (1999) coastal structures model was used to estimate the regional economic impact of sea level rise in the United States. This model estimates the cost of rising sea level over time, based on comparing the cost of protecting the coastline from inundation with the benefits of this protection. The benefits of protection are the opportunity cost of coastal property that would be abandoned if inundated. The true opportunity cost of abandoning coastal property should reflect the projected value of property at the future time of inundation as well as adaptive measures that would be taken to minimize property loss. The model then includes a representation of the future trajectory of property value for land and property threatened by inundation. In addition, it incorporates two types of adaptive measures. First, the value of land lost to inundation is represented by the value of land located inland from the ocean. At the point of inundation, any price gradient associated with closer proximity to the ocean simply migrates inland, so that in most cases the real loss is best represented by the value of inland property. Second, if there is sufficient foresight, structure value should depreciate in the face of a growing risk of inundation. The depreciation mitigates the losses associated with inundation. Full depreciation of structure value in anticipation of inundation is reflected in the 'perfect foresight' model runs. Alternatively, an efficient process of depreciation could be hampered because there is not enough time, the risk communication is ineffective, risk perceptions are faulty, or owners incorrectly expect that their land will be protected by public action. The 'no foresight' model runs reflect no depreciation of the structure value, effectively bounding the impact of this adaptive measure.

The cost of coastal protection is based on two protection alternatives: hard structure armoring through the construction of dykes, seawalls, or bulkheads; and the placement of sand on the beach, often referred to as beach nourishment. The capital cost of hard structures of $935 per linear foot (in year 2000 dollars) was derived from a review of published studies; the value represents a central estimate from those studies. We model maintenance costs as a percentage of construction costs, again based on estimates reported in published studies. We chose a 4 percent maintenance cost per year as the central estimate, but 10 percent was used for hard structures that might be built along coastline directly open to the ocean. The capital costs also reflect differences in the cost of building structures of different heights. For example, because the base of the required protective structure expands with its height, the structure necessary to protect property from a 1 m rise costs more than twice as much to construct as that necessary to protect the property from a 0.5 m rise. Finally, costs to nourish beaches were modeled using estimates of the requisite volume to nourish the full beach profile at a rate that matched the relative sea level rise and regional estimates of the price of sand. Beach nourishment is assumed to be necessary starting in 2010 (the first decadal estimate in our analysis), and is assumed to be effective as long as the sea level rise does not exceed 30.5 cm. Beyond that threshold, we assume that a hard structure constructed at the back of the beach is necessary to ensure protection of interior property.

The model simulates the protect–abandon decision as a dynamic cost–benefit comparison through time. Using a decadal time-step, the model calculates the net benefits of protection at the point when inundation is imminent. If net benefits are positive, the capital costs for a protective structure are incurred just before inundation, and maintenance costs are incurred for all subsequent years. If net benefits are negative, the land is abandoned and the opportunity costs of losing both the land and structure value are incurred. In theory, the dynamic nature of the model allows for situations where it might be reasonable to protect for some period of time but, as property values change over time, the benefit of protection could fall to a point where it is exceeded by the present value of the future stream of maintenance costs. In fact, the increasing trend in property values over time, associated mainly with increasing per capita income, ensures that once protection is calculated to have net benefits, continued protection remains the optimal course. The projected upward trend in development value reflects historical patterns of coastal development over the three decades before 1990, and is an important factor in accurately assessing future impacts in the coastal zone.

For this analysis, we used the model to generate regional sea level rise cost estimates for four scenarios of future eustatic sea level rise between 1990 and 2100: 33, 50, 67, and 100 cm. We assume a scenario involving a steady increase in the rate of sea level rise, which appears to best match the trajectory of sea level rise anticipated in IPCC (2001). For each of these four sea level rise scenarios, we generated two sets of economic impact estimates, one assuming perfect foresight and the other assuming no foresight.

Along with the rate of sea level rise (33, 50, 67, 100 cm) and the assumption about foresight, a host of other factors influence the economic impact of sea level rise. Factors such as growth in state GDP, state population growth, and discount rate are model inputs. We relied on the socioeconomic scenarios described in Chapter 2 for these values. The model also takes as input site-specific economic parameters such as the value of land and structures and the required length of protective structure necessary.

The model does not directly estimate regional impacts of sea level rise. Instead, it generates site-specific cost estimates for a given set of inputs. We scale these site-specific model results to the full state using scaling procedures that make use of shoreline length data and sea level rise vulnerability estimates. As part of this project, we added new sites using the procedure established in Yohe et al. (1999) and Neumann and Livesay (2001), for a total of seven California sites. The sites chosen were designed to yield a base of sample sites from which we could extrapolate effects for the full California coast (see Figure 13.1 for a map of the site locations). The four new sites chosen were Imperial Beach, which includes the area at the southern tip of San Diego Bay, with both open ocean and bayside frontage; Año Nuevo and the cliff site south of San Francisco; Palo Alto, in San Francisco Bay, which includes several low-lying residential areas; and Ferndale, along Humboldt Bay in northern California. At three of the four new sites, inundation mapping indicates that developed areas are vulnerable (the exception is Año Nuevo). In addition, the sites represent a diverse set of coastal topography and property value characteristics, and therefore a diverse set of coastal vulnerabilities to sea level rise, from which we can develop a statewide estimate of economic impact.

Under the four sea level rise scenarios and both foresight assumptions, we estimate the impact of sea level rise on these seven individual coastal sites. These seven sites, presented in Figure 13.1, were originally drawn from a larger set of sites chosen by Park et al. (1989),[2] to serve as a national sample for assessing the economic damage induced by sea level rise in the United States (see Yohe, 1990). For modeling purposes, each site is divided into 500 m by 500 m grid cells.[3] Then, based on estimates of the timing of the inundation for each grid cell found in the Park et al. (1989) work, the model estimates the economic cost of sea level rise at each site. The timing of inundation proves to be crucial in our estimates of the current economic impact, because we assume that protective seawalls or dykes need not be built until just before inundation (although beach nourishment must begin immediately to be effective). The current value of the impact estimates reflects discounting of costs back to the start of the study period. Reliance on the Park et al. (1989) sites provides more precise estimates of the timing of inundation, although it also limits our procedure to those sites modeled by the Park team (see the attachment to Appendix XIII at http://www.energy.ca.gov/reports/2003-10-31_500-03-058CF_A13.PDF).

We aggregate the site-specific results to the state level using a series of scaling factors. We base our scaling factors on sea level rise vulnerability

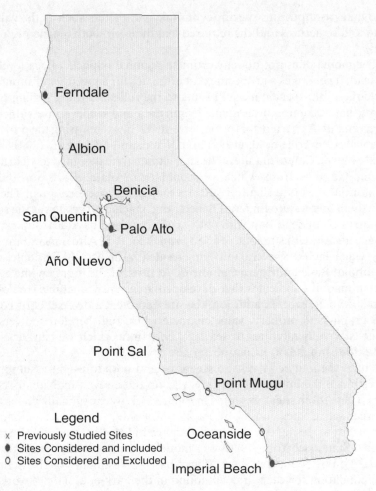

Figure 13.1   Sea level rise candidate sites in California

estimates developed by the US Geological Survey (USGS), including the coastal slope and coastal vulnerability index (CVI) (see Thieler and Hammer-Klose, 2000). Specifically, we develop scale factors by comparing vulnerability estimates for the sites to average estimates for the state. USGS develops its CVI estimates based on a range of vulnerability factors (including tidal range, wave height, coastal slope, shoreline change, geo-morphology, and historical rate of sea level rise). As a result, we believe that the CVI estimate provides the broadest indicator of vulnerability. Because CVI data are unavailable for most of the San Francisco Bay, we also apply a concurrent scaling approach for this area. We assume that the cost per mile for the Palo Alto site is equal to the average per-mile cost for the entire bay.

For comparison purposes, we also scale costs to the full state based solely on shoreline length of the sites versus the length for the entire state. This simplistic alternative makes the strong assumption that the seven sites we selected are representative of all California sites without any weighting to adjust for characteristics that should affect sea level rise vulnerability. For this scaling approach, we determine the average cost per mile for all seven sites (weighted by the sites' shoreline length) and multiply by the total state shoreline length. It is not likely, however, that these sites are a representative sample – we calculated these results only for comparison purposes.

For the seven sites used, developed areas represent only a portion of the total land for which inundation is expected. As a result, we most likely underestimated the amount of developed land at risk along the highly developed shoreline between Santa Monica and San Clemente in the Los Angeles area. Therefore, because this area will require significantly greater protection and beach nourishment than the seven sites, the vulnerability-based scaling approach most likely underestimates protection costs. To address this potential bias, we also developed a second set of estimates using each of the four scaling methods discussed previously by estimating the cost of protecting the entire Los Angeles area with beach nourishment and protection costs. We use the beach nourishment and armoring costs for the 1.5 miles of developed shoreline in the Imperial Beach site and apply these costs to the 87-mile Los Angeles area shoreline. This approach may overestimate costs, because it is not clear that all shoreline segments in the Los Angeles area will require beach nourishment and seawall protection. For example, we expect that commercial port facilities are already armored shorelines; the incremental cost of sea level rise is to raise the maintenance cost of these armoring structures, and perhaps add construction and rebuild costs as the armoring wears out, but clearly no beach nourishment is required. As a result, we believe that the 'LA adjustment' procedure provides a better estimate than the basic scaling results, but may nonetheless overstate actual protection costs.

13.3   RESULTS

Table 13.1 reports the economic cost of sea level rise under all four scenarios at two discount rates (3 percent and 5 percent) for each of the seven individual sites and for the state as a whole using several different aggregation schemes. The values are present values and reflect the sum of the

Table 13.1 Present value of cost estimates for coastal protection strategy from 2000 to 2100 under four sea level rise scenarios (millions of year 2000 dollars)

	33 cm SLR scenario		50 cm SLR scenario		67 cm SLR scenario		100 cm SLR scenario	
	3% dr	5% dr	3% dr	5% dr	3% dr	5% dr	3% dr	5% dr
Site estimates								
Site name								
Albion	Zero inundation							
Año Nuevo	Zero inundation							
Ferndale	0.127	0.026	0.741	0.187	1.757	0.513	4.922	1.496
Imperial Beach[a]	0.329	0.051	0.626	0.205	1.178	0.344	3.455	0.884
Palo Alto[b]	2.484	1.125	8.015	3.832	14.589	7.064	34.744	16.715
— Newark	1.882	0.877	5.986	2.933	10.676	5.257	25.288	12.368
— Redwood Shores	0.604	0.248	2.029	0.899	3.914	1.807	9.456	4.347
Point Sal	Zero inundation							
San Quentin	0.278	0.122	0.994	0.409	2.059	0.849	5.509	2.176
Full state estimates								
Scaling approach								
CVI	$25	$10	$82	$36	$160	$67	$407	$168
Slope	$22	$9	$74	$31	$142	$60	$358	$148
Shore length	$27	$10	$90	$37	$175	$74	$445	$182
CVI with LA adjustment	$47	$12	$117	$46	$224	$87	$597	$216
Slope with LA adjustment	$41	$11	$108	$42	$208	$80	$551	$198
Shore length with LA adjustment	$46	$14	$123	$49	$238	$92	$635	$229

Notes:
All estimates assume a rate of 4% for variable costs of protection, unless otherwise specified. SLR is sea level rise; dr is discount rate.
a. An open ocean site that employs a beach nourishment strategy and a 10% variable protection cost.
b. Estimates are the sum of results for subsites analyzed at that location.

discounted costs for the entire century. The values show that the total esti-mated economic impact of a 100 cm sea level rise in California varies from $148 million (based on coastal slope scaling and a 5 percent discount rate) to $635 million (based on shore length scaling, a 3 percent discount rate, and the LA adjustment). Because the impacts are greatest toward the end of the century, the discount rate plays a large role in these present values. Comparing across scenarios, the expected economic impact of sea level rise increases sharply with more rapid sea level rise trajectories. As illustrated in Figure 13.2, the rate of sea level rise has a very large impact on costs. The estimates presented in Table 13.1 and Figure 13.2 suggest that the scaling procedure has a major effect on the results if the seas rise rapidly (the 100 cm case). In this case, the method of extending results to the Los Angeles region, in particular, has a major effect on the results. However, with the slower sea level rise scenarios, the different scaling approaches yield very similar results. Note that the results presented in Table 13.1 reflect a no fore-sight assumption – foresight has only a modest effect on the results in California because a protection response is implied for all but the San Quentin site. In our model, foresight has no effect on the marginal protec-tion costs, affecting only the costs of abandonment. The almost uniform finding at these California sites that protection strategies dominate retreat stands in contrast to our prior national results, where retreat or partial retreat was an efficient response at approximately one-third of the sample sites nationwide (Yohe et al., 1999). Our finding that protection is an eco-nomically efficient response at most developed sites in California reflects the relatively high economic values for California coastal properties. As a result, assumptions about whether or not property owners will depreciate their structures before inundation are not relevant to California and have no effect at the sample sites chosen.

Although the present value of costs provides a good overview of the overall burden caused by sea level rise, it is also insightful to examine the timing of these costs. Figures 13.3 and 13.4 and Table 13.2 illustrate the annual costs of sea level rise using CVI scaling under all four scenarios, cal-culated without and with the LA adjustment, respectively. Annual costs increase rapidly over time to keep pace with the increasing trend in sea level rise. This policy reflects the 'just-in-time' need for capital investment in pro-tection. Costs up to 2040 largely represent the ongoing cost of beach nour-ishment, which begin in 1990 and grow as increasing volumes of sand are needed to maintain a constant beach profile. The lump sum investments for hard structures that begin in 2040 are followed by ongoing expenditures required for operation and maintenance, as well as ongoing beach nourish-ment at open ocean sites. In prior work, the trajectory of costs was highly dependent on the pattern of inundation at just a few sites. Our current

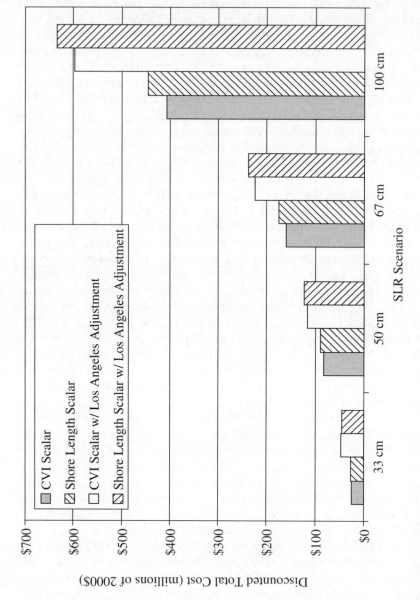

Figure 13.2 California's economic cost of sea level rise

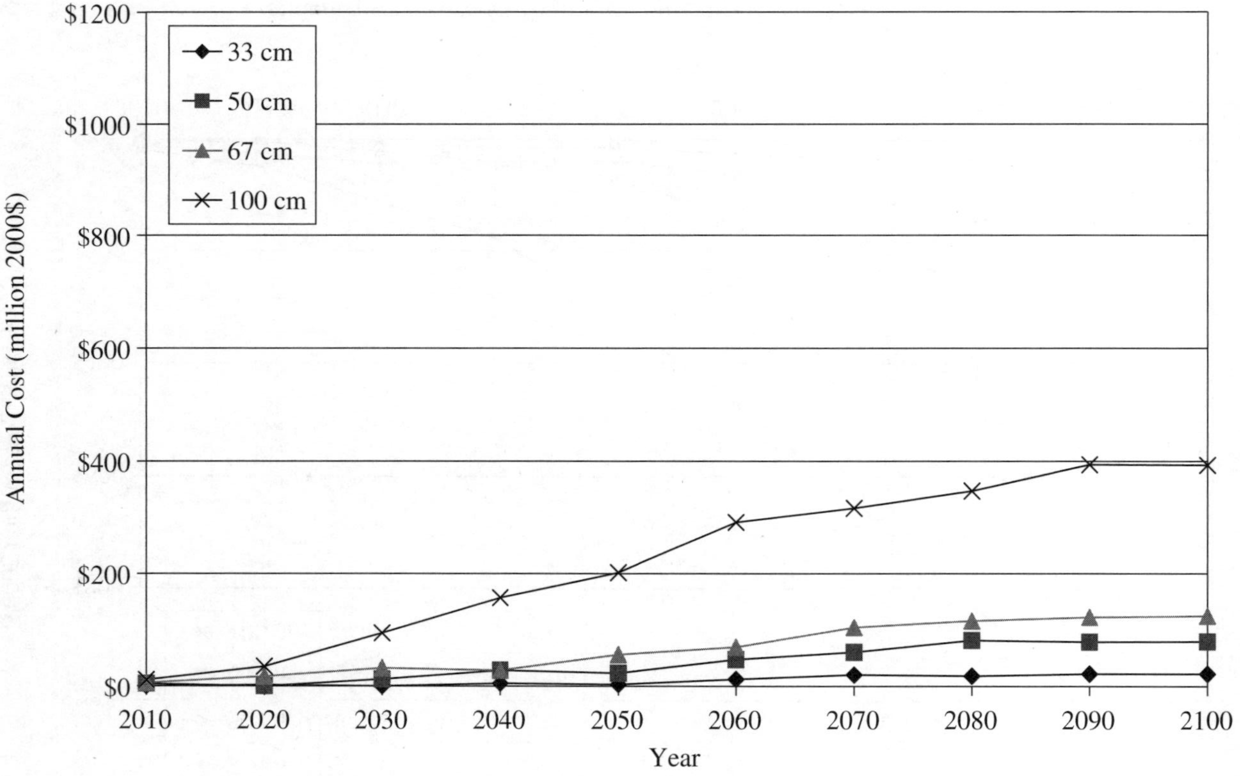

Figure 13.3   California's transient costs for sea level rise without Los Angeles adjustment

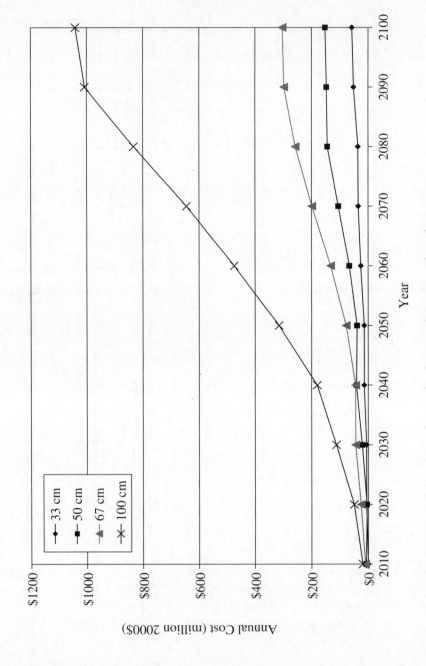

Figure 13.4 California's transient costs for sea level rise with Los Angeles adjustment

Table 13.2   California decadal transient costs (millions of undiscounted 2000 dollars)

Year	CVI without LA adjustment				Year	CVI with LA adjustment			
	33 cm	50 cm	67 cm	100 cm		33 cm	50 cm	67 cm	100 cm
2010	0	4	6	11	2010	0	7	10	19
2020	2	0	17	35	2020	6	4	25	49
2030	0	12	32	94	2030	6	21	44	112
2040	5	28	28	157	2040	14	41	44	180
2050	3	23	56	202	2050	14	39	77	315
2060	12	47	69	292	2060	26	66	132	474
2070	20	60	104	317	2070	35	105	199	645
2080	18	81	116	348	2080	36	144	257	833
2090	22	78	122	394	2090	51	147	297	1 007
2100	22	79	124	393	2100	57	152	303	1 040

results reflect a greater number of sites, yielding a smoother trajectory over time, although the results are still largely dependent on the results for the Imperial Beach site, which drives results for most of the Southern California region. These results suggest that capital and operating costs for hard structures and beach nourishment could steadily climb to approximately $1 billion (in year 2000 dollars) per decade by 2100.

Table 13.3 illustrates how the costs of sea level protection are distributed within California. San Francisco Bay and the South Coast (which includes the Los Angeles area) incur the majority of costs. In contrast, the less densely developed and more sheltered coast along the northern and middle portion of the state accounts for less than 25 percent of the total costs. The low estimate for the South Coast area reflects the basic scaling approaches; the high estimate reflects the basic scaling approach with the LA adjustment.

13.4   CONCLUSIONS

Our work provides a first approximation of the economic implications of sea level rise for coastal structures in California, including estimates of costs through time. The extension of previous work to reflect the results at a greater number of sample sites improves those earlier estimates, but limitations in the readily available inundation data result in statewide estimates that remain uncertain to within roughly a factor of two. To the extent that

*Table 13.3   California's regional economic cost of sea level rise (millions
of year 2000 dollars)*

CVI scaling

Region	33 cm	50 cm	67 cm	100 cm
North Coast	2	8	16	46
San Francisco	13	43	79	189
Mid Coast	2	6	12	33
South Coast	3–18	10–37	21–73	57–210
Total	20–35	66–94	128–180	326–479

Shorelength scaling

Region	33 cm	50 cm	67 cm	100 cm
North Coast	2	8	16	46
San Francisco	13	43	79	189
Mid Coast	2	6	12	33
South Coast	3–18	10–37	21–73	57–210
Total	20–35	66–94	128–180	326–479

policy development applications at the state level require more precise estimates, additional effort should be devoted to understanding the expected
inundation pattern in the Los Angeles area. Our adjusted estimates provide
a first approximation of protection costs for this region, but do not yet
reflect careful modeling of the inundation pattern or site-specific consider-
ations of the response strategies for commercial/port versus urban/resi-
dential versus lower density residential development. Our estimates with
and without the LA adjustment most likely bracket the estimates that could
result from more careful modeling of response options that vary by land
use type, because the adjustment assumes that the entire LA region adopts
the most expensive response option, beach nourishment followed by
seawall construction at the back of the nourished beach. In addition, the
current estimates do not reflect potential impacts of sea level rise in some
estuarine or brackish areas of San Francisco Bay, particularly the San
Joaquin Delta, which are currently vulnerable to riparian flooding and may
face an increased risk of flooding as the bay level rises.

Three other refinements may also yield interesting insights that can guide
efforts to increase California's capacity to adapt to sea level rise. First, we
did not examine impacts to wetland sites that we know will be inundated
by rising seas. These sites provide benefits to many citizens by supporting

wildlife, recreation opportunities, and open space. Unfortunately, the value of these services is difficult to quantify. Nonetheless, land use planning in the coastal zone can play an important role in adapting these sites to sea level rise. Forward-looking land-use planning around wetland areas could preserve dryland landward of vulnerable wetlands. As the seas rise, the wetland could then migrate inland, maintaining at least a portion of the ecological value of the wetland. California should also look carefully at coastal land that is currently used for military purposes. By protecting some of this land from future development, the state may be able to create some coastal zone ecosystems that will be lost elsewhere. Finally, the state should conduct systematic analyses quantifying the value of coastal wetlands and estimating the viability and migration capacity of important wetland functions in the face of rising seas.

Second, ongoing work to estimate future land use patterns in California could serve as the basis for incorporating a dynamic land use input to the economic model employed here. The greatest value of such work would be for currently undeveloped or less-developed sites, and potentially at current urban sites where the possibility of economic decline or adaptive redevelopment is evident. In addition, the possibility of sea level rise vulnerabilities affecting private and public investment in coastal infrastructure suggests that projections of future land use in the coastal zone could be substantially influenced by processes of learning and adaptive responses to this risk. By integrating this modeling approach in a GIS framework, users could examine the economic impacts of development and shoreline armoring decisions. Such a framework would facilitate the use of finer resolution data and local-level analyses.

Third, erosion and storms also affect California's coastal area by damaging structures and in some cases leading to earlier protection decisions. For example, El Niño events have damaged the state's energy infrastructure and cliffside homes. Thus, for the extensive cliff areas along California's northern coastline, it would be useful to identify the rate of erosion and the resulting cost of either stabilizing the cliffs or allowing structures to fall into the ocean. For storm-related flooding, it would be useful to estimate the cost of storm damage as a function of height above sea level. This would provide a measure not only of inundation but also of storm damage. The resulting values would inform decision makers about when it would be best to construct protective structures as sea levels rise. The timing of such responses is a key factor in estimating the present value of coastal planning decisions.

One of the key results of this chapter is the importance of timing. As stated earlier, most of the costs of sea level rise in California are likely to be protection costs along the coast. Furthermore, in our analysis we assumed that the protection actions will be taken at the optimal time, just before properties are inundated. If protective actions are taken before the optimal time, the capital and maintenance costs would be shifted earlier, making them more expensive in present value terms. To illustrate the importance of timing, if protective seawalls were constructed ten years sooner than we estimate, their discounted capital costs would increase by 35 percent to 63 percent, depending on the choice of discount rate; if they were constructed 20 years too soon, discounted costs would increase by 81 percent to 165 percent. If protection actions are taken later than the optimal time, there could be property damage that, according to our calculations, would exceed the cost of protection, again raising costs relative to our estimates. The likelihood that protection is constructed at the optimal time is lessened by the fact that concerted action is required of multiple property owners – each landowner cannot simply erect his or her own seawall, because without a similar response from neighboring properties the water will simply go around a single property wall. However, the key point is that the timing of response to sea level rise is critical to the estimate of its cost. We urge that sea level rise rates continue to be monitored and that state and local planners update their measures of property development. With up-to-date information over time, decision makers can ensure that their responses carefully match the threat of inundation.

ACKNOWLEDGMENTS

The authors thank Jane Leber Herr, who collected much of the site data in California, and Jennifer Kassakian, who generated the necessary sea level model data for new sites, for their able research assistance.

NOTES

1. Because the Park et al. (1989) model examines inundation patterns for Calfornia sites along the Pacific Ocean and San Francisco Bay area, we were unable to examine economic impacts along low-lying portions of the San Joaquin and Sacramento rivers.
2. The Park et al. (1989) site selection is described as follows: 'Ninety-three sites were chosen, using an unbiased systematic sampling of US Geological Survey topographical maps at a scale of 1:24 000. Starting with the easternmost quadrangle in Maine and restricting the choice to those maps that included some part of the coast, every 15th quadrangle was picked as the center of a site consisting of one to four quadrangles.' As described in the text, Yohe (1990) interprets the results of the Park et al. sampling to yield a roughly one-tenth sample of 30-minute cells, provided by the USGS. For the initial Yohe et al. (1999) work, and for subsequent economic modeling, a subsample of 30 sites was chosen from the 93 sites in the Park et al. team's work.
3. At five sites, grid cells are 250 m by 250 m; this finer grid was used at Albion, Año Nuevo, Imperial Beach, Point Sal, and San Quentin.

REFERENCES

California Coastal Commission. 2001. Overview of Sea Level Rise and Some Implications for Coastal California. California Coastal Commission, San Francisco.

Gleick, P.H. and E.P. Maurer. 1990. *Assessing the Costs of Adapting to Sea-level Rise: A Case Study of San Francisco Bay*. Pacific Institute for Studies in Development, Environment, and Security, Berkeley.

IPCC. 2001. *Climate Change 2001: The Scientific Basis. Contribution of Working Group I to the Third Assessment Report of the Intergovernmental Panel on Climate Change (IPCC)*. Cambridge University Press, Cambridge, UK.

Mendelsohn, R. 2001. Adaptation. In *Global Warming and the American Economy*, R. Mendelsohn (ed.). Edward Elgar Publishing, Cheltenham, UK and Northampton, MA, pp. 167–86.

Neumann, J. and N. Livesay. 2001. Coastal structures: Dynamic economic modeling. Chapter 6 in *Global Warming and the American Economy*, R. Mendelsohn (ed.). Edward Elgar Publishing, Cheltenham, UK and Northampton, MA, pp. 132–48.

Park, R., M. Trehan, P. Mausel, and R. Howe. 1989. The Effects of Sea Level Rise on US Coastal Wetlands. In *The Potential Effects of Global Climate Change on the United States: Appendix B – Sea Level Rise*, J.B. Smith and D.A. Tirpak (eds). Report to Congress. US Environmental Protection Agency, Washington, DC.

Thieler, E. and E. Hammer-Klose. 2000. National Assessment of Coastal Vulnerability to Sea-Level Rise: Preliminary Results for the US Pacific Coast. US Geological Survey, Woods Hole, MA. Available at http://pubs.usgs.gov/openfile/of00-178/.

Titus, J.G., R.A. Park, S.P. Leatherman, J.R. Weggel, M.S. Greene, P.W. Mausel, S. Brown, C. Gaunt, M. Trehan, and G. Yohe. 1991. Greenhouse effect and sea level rise: The cost of holding back the sea. *Coastal Management* **19**: 171–204.

Yohe, G. 1990. The cost of not holding back the sea: Toward a national sample of economic vulnerability. *Coastal Management* **18**: 403–32.

Yohe, G., J.E. Neumann, and P. Marshall. 1999. The economic damage induced by sea level rise in the United States. In *The Impact of Climate Change on the United States Economy*, R. Mendelsohn and J.E. Neumann (eds). Cambridge University Press, Cambridge, UK, pp.178–208.

Yohe, G., J. Neumann, P. Marshall, and H. Ameden. 1996. The economic cost of greenhouse induced sea level rise for developed property in the United States. *Climatic Change* **32**: 387–410.

14. Conclusion
Robert Mendelsohn and Joel B. Smith

14.1 INTRODUCTION

This book investigates the impact of climate change on California. It explores what methods can be applied at the regional level, and furthers the science of assessing regional impacts of climate change. The book (1) develops detailed scenarios of future changes in population, income, and land use; (2) conducts in-depth and integrated quantitative analysis of climate change impacts on natural systems (ecology, land, and hydrology); (3) explores the impacts of these physical changes at a regional scale on important sectors (water resources, agriculture, and others); and (4) analyzes in depth the potential for economically efficient adaptation to ameliorate adverse impacts of climate change or take advantage of positive impacts.

Regional studies can explore effects at a much finer grain than national and global economic impact studies, and the book explores what can be learned from this fine detail. For example, the biodiversity study (Chapter 6) examines small valleys within neighborhoods and discovers possible refugia for threatened plants that would not have been visible at much larger scales. The coastal study (Chapter 13) models the individual beaches along the California coastline and provides a more accurate measure of the impact of sea level rise. The timber (Chapter 7), water (Chapter 10), agriculture (Chapter 11), and energy (Chapter 12) studies all provide measures of how climate will impact different parts of the state. Earlier, more aggregated studies could not see these more refined distributional impacts.

One important strength of the research in this book is the careful integration across both the natural sciences and economics. For example, a common baseline was developed for the entire study of land, water, and economic uses in the future. Two climate scenarios, the Hadley and the PCM, were then evaluated across all the sectors. The outputs of many studies then became the inputs of others. For example, the ecosystem effects reported in Chapter 5 were used to model biodiversity in Chapter 6 and timber in Chapter 7. The outputs of the runoff and crop yield modeling in Chapters 8 and 9 were integrated into Chapter 10 on water and Chapter 11 on agriculture. The study was able to forge a solid link between physical changes such as runoff and societal consequences such as impacts on agriculture.

The study also addressed an important methodological question concerning whether impact models should take prices as given or whether they should predict prices endogenously. Regional studies must be careful when assuming prices are either exogenous or endogenous. For systems that lie totally within the region, prices are generally endogenous. For example, in California's hydrological system, the value of water is determined strictly by the state's demand and supply. In this case, water prices are endogenous; they will change with climate. Chapter 10 consequently estimates how the shadow value of water could change. California may also dominate other sectors such as the production of some fruit and nuts for US consumption. Such systems could also be modeled with endogenous prices. However, in many sectors, the region is only a small player. In those markets, prices are exogenous to what happens in the region. Chapter 11 consequently takes the prices of grains as exogenous. Ideally, a regional study will have a more global analysis that can provide price forecasts for specific climate scenarios such as is done in Chapter 7 on timber markets. Otherwise, prices must be assumed to be exogenous to the study (as was done in Chapter 12 on energy). That is, regional analyses must be careful not to assume that prices depend only on what happens in the region.

Another important innovation in this book is that the studies examine the interaction between climate change and changes in baseline socio-economic conditions. In this study we estimate climate impacts out to 2100. To understand what these future changes in climate might do to society, it is important to imagine what the society might look like in that future time. For example, in a state such as California, the population is expected to grow, primarily from immigration. The economy is also expected to grow both in absolute and per capita terms. How much population and the economy will grow over such a long time period, however, is hard to predict. The study consequently looks at a slow and fast growth scenario to determine whether the speed of growth affects the magnitude of the climate impacts.

To measure the impact of climate change on a region, it is important to examine a range of climate changes. Although scientists are convinced that increases in greenhouse gases will lead to higher temperatures and more precipitation globally (Houghton et al., 2001), the magnitude of the warming and the changes to precipitation in a region such as California are highly uncertain. To reflect the uncertainties about climate change in California, it is therefore important to examine the range of plausible scenarios that could occur. In the case of California, it is critical to present both moderate and large changes in temperature and both wet and dry scenarios. Any study that just presents one or two hot and dry scenarios or warm and wet scenarios alone would give the false impression that this is the only outcome possible from global warming and might encourage a false sense of prediction about future changes.

14.2 RESULTS

The book contains not just methodological advances but also empirical results for California. The results include predictions of physical changes in systems such as movements in ecosystems and changes in spring runoff. More important, perhaps, the book has measured the sensitivity of key sectors to climate. This sensitivity is then used to predict how climate change would affect society. The resulting analyses measure the economic impacts to the agriculture, coastal, energy, timber, and water sectors in dollars (see Table 14.1). However, not everything can be measured in dollars. The book also measures the impact of climate change on biodiversity.

Table 14.1   The estimated annual market impacts of climate change in 2100 ($ millions)

Scenario/study	HadCM2	PCM	+3°C, 0% P	+3°C, +18% P	+5°C, 0% P	+5°C, +30% P
Urban water use	3	−87				
Hydropower	85	−46				
Operating costs of water system	237	−147				
Agriculture	15	−1100				
Direct timber	76	86	−124	−93	−159	−86
Timber price change-supply	−192	−265	−133	−129	−117	−123
Timber price consumer	1500	1500	1500	1500	1500	1500
Energy	−11 200	−6900	−8000	−6700	−17 800	−18 900
Coastal (sea level rise scenarios)	−30 (0.67 m)	−6 (0.33 m)	−15 (0.5 m)	−30 (0.67 m)	−104 (1 m)	−104 (1 m)
Total market impacts	−9779	−7016	. . .	. . .	. . .	. . .

Note: Positive numbers are benefits and negative numbers are damages. Energy estimates assume a price increase. Sea level rise scenarios have been assumed given predicted temperature change. The direct timber effects capture the impacts of climate change on forests in California. The timber price effects capture how price changes from around the world would affect timber suppliers and consumers. Values were not calculated for the incremental scenarios in some sectors.

Ecosystems, Biodiversity, and Timber

Terrestrial ecosystems

The ecosystem model (Chapter 5) predicts that under all climate change scenarios, forests and other types of vegetation will migrate to higher elevations as warmer temperatures make those areas more suitable for survival. The area of alpine and subalpine forests will shrink as evergreen continental forests and shrublands migrate to higher altitudes with higher temperatures. If climates get wetter, forests would expand in northern California and grasslands would expand in southern California. If climate gets drier, grasslands would expand across the state. Both wetter and drier scenarios result in increased carbon in vegetation, meaning there would be an increase in biomass. Interestingly, the frequency and the size of fires would increase under wet and dry scenarios, but not until the latter part of the century. The drier scenarios result in more frequent fires and more area consumed by fires. The wetter scenarios result in fewer fires, but the fires would have greater intensity because there would be more fuel (vegetation) during occasional dry periods.

Diversity of terrestrial communities

The biodiversity study (Chapter 6) combines the estimated urbanization patterns (Chapter 3) with the estimated terrestrial ecosystem changes (Chapter 5) to examine implications for terrestrial biodiversity. The study found that climate change would change the distribution of vegetation types across the state more than urbanization. Over time, urbanization is expected to continue reducing diversity of vegetation communities. The effect is mainly on Mediterranean shrublands and grasslands between Los Angeles and San Diego. The impacts of climate changes are more complex. Some ecosystems such as arctic and Mediterranean shrublands are estimated to decrease under wet and dry scenarios. Some forested ecosystems and some grasslands (C_4) could increase. On the whole, with warm and wet climates, there would be little change or even a slight increase in the diversity of California's vegetation communities. However, with a warm and dry climate, community diversity would be reduced.

The biodiversity study also examined the combined impact of urbanization and climate change on a threatened and valuable habitat along California's coast, coastal sage scrub (CSS). Development has already reduced CSS by 90 percent, and projected development during this century could reduce the current CSS habitat by up to another 20 percent. Under all the climate change scenarios, only a small fraction of CSS would persist, and more than half of these potentially surviving areas would be threatened by development. This preliminary analysis identifies specific areas (east of San Diego) that are not yet developed where the model estimates that CSS could survive climate change. These areas may be prime locations for future conservation efforts for CSS. By looking far into the future, the study identifies areas that should be protected from new development today.

Timber

The timber study (Chapter 7) combines the estimated changes in location and productivity of California vegetation over time from Chapter 5 with a timber production model to assess the potential impacts of climate change on timber markets. This study focuses on softwoods on privately owned lands, assuming that little timber on public lands would be harvested. Commercial timberlands could expand in northern California but are projected largely to shrink in the rest of the state, especially toward the end of the century. However, increased CO_2 concentrations in the atmosphere are predicted to increase forest productivity in the near future, particularly under warmer and wetter conditions. The net result of these changes would be an increase in supply of California-grown timber throughout the next several decades but a possible decline closer to the end of the century.

The timber study also examined how sensitive the sector was to changes in world prices. Timber price changes have a large impact on California. If global timber prices fall, as predicted in published studies, producers face losses, even in the near term, because their gross revenues will fall. Consumers, on the other hand, will derive large benefits from lower prices. Under such scenarios, the net economic effect to Californians is beneficial because California consumes far more timber products than it produces. However, within the state, there would likely be winners and losers. Specifically, consumers throughout the state, but especially those in urban areas, would benefit, whereas mountain and northern communities that are more dependent on forest production would see relatively more losses.

Runoff, Water Supply, Crop Yields, and Agricultural Production

Runoff

The runoff study (Chapter 8) estimated changes in snowpack, snowmelt, and runoff in six key basins: Smith River, Sacramento River, Feather River, American River, Merced River, and Kings River. The study predicts that higher temperatures will melt snowpacks earlier in the year and cause more precipitation to fall as rain rather than snow. This results in substantial reductions in the size of the snowpack under the relatively dry PCM scenario and slight decreases under the wet Hadley scenario. An additional consequence of higher temperatures is earlier and higher peak flows, which in turn could result in an increased risk of flooding. Even under scenarios with reduced precipitation, such as PCM, peak flows were estimated to increase. Those peak flows were predicted to increase substantially under the wet scenarios. Changes in total annual runoff varied with changes in precipitation. Given these changes in seasonal runoff by basin, it is possible to calculate state-wide changes in runoff. The PCM scenario results in a 26 percent reduction in runoff from the mountains and the Hadley scenario results in a runoff increase of 77 percent. The incremental scenarios lie between these outcomes, ranging from virtually no change to increases up to 28 percent.

Water supply
The water model in Chapter 10 combines the results of estimated changes in runoff from snowpack, surface water, groundwater inflow, and reservoir evaporation to estimate effects on total raw water availability. The changes in total raw water availability range from a 12 percent increase under Hadley[1] to a 25 percent decrease under PCM. The incremental scenarios result in either little change or in reductions. The water model can move water from one location to another and thereby reduce some of the impact of runoff changes.

In addition to capturing changes in supply, the water model also reflects changes in the demand for water. Baseline changes from population growth, for example, are presumed to increase urban demand for water by two-thirds, thereby reducing water use for agriculture in California by about 10 percent. The model also captures changes in water demand caused by climate change. For example, warming changes evapotranspiration, which in turn changes the amount of water that crops need per hectare (Chapter 9).

Combining changes in demand and supply, the water model (Chapter 10) allocates water to its highest valued use, that is, where it can be used most efficiently. In essence, the model takes water from low-valued users and gives it to higher-valued users. In a wet scenario, such as Hadley, demand increases but supply increases even more, resulting in water prices falling. Under Hadley, both urban and agriculture users would get slightly more water than they would with no climate change. Under the PCM scenario, however, supply falls dramatically and the price or scarcity of water increases. In this case, urban deliveries would be reduced by 100 000 acre-feet per year, or about 1 percent, but agricultural deliveries would be reduced by 8.8 million acre-feet, or about 35 percent. The agriculture sector would be the hardest hit in dry scenarios. Water scarcity is not expected to affect southern California unless there is a change in the water supply from the Colorado River (this study assumed no change in water supplies from the Colorado River).

Hydropower production in the California system would also be affected by climate change. Just looking at the dams of the major water supply reservoirs, annual hydropower revenue would fall 30 percent, or $48 million by 2100 with the PCM scenario. This loss does not include the impacts of climate change on the other hydropower plants in California. However, it is likely that they too would face a 30 percent reduction in revenue. In contrast, with the wet Hadley scenario, annual hydropower production is projected to increase by 52 percent, or $85 million by 2100. Whether future climates are wetter or drier will have large consequences for hydropower effects.

Crop yields
The study in Chapter 9 begins by modeling the effects of climate change on crop yields. Physical process models exist for only a few of the crops grown in the state, so cross-sectional models were estimated. Statistical relationships were estimated that link county-level crop yields to monthly growing season temperatures and precipitation. Carbon fertilization effects from higher CO_2 levels were calculated from the results of plant growth models. The combination of higher temperatures and CO_2 fertilization tends to result in increased crop yields in northern and coastal regions and decreased crop yields in the San Joaquin and desert areas. The demand for irrigation water is estimated to rise by a few percent. The chapter also found that technology improvements could substantially increase yields with changes far greater than the potential impacts of climate change. The study assumed that technical change would not change the relationship between yields and climate. However, if climate change reduces crop yields proportionally and technical change increases the absolute magnitude of crop yields, the absolute magnitude of the climate change impacts will be larger.

Agricultural production
The changes in yield, available farmland, water use by crop, and water supplies for agriculture were all entered into an economic model for agriculture in Chapter 11. This model, the Statewide Water Agricultural Production (SWAP) model, maximizes each region's total net returns from agricultural production given the resource constraints. The overall effects of the potential changes in water supply on the value of statewide agricultural production are relatively small, although some counties could be more severely affected. The wet Hadley scenario results in only a small increase in statewide production and a small increase in net revenue. The approximate 25 percent decrease in water supplies for agriculture under the PCM scenario results in a 6 percent decrease in agricultural income, or about $1 billion. The reason that the percentage change in income is so much smaller than the percentage change in water deliveries is that the reductions in crop

acreage are concentrated mainly in low-value, high water-using crops such as alfalfa. The loss of marginal quantities of water leads to the loss of only low-valued production.

While the overall effects on agriculture appear to be relatively small, these effects are not felt equally across the state. Some regions get slightly more water even under PCM, while other regions have reductions in deliveries of about three-fourths and reductions in income of one-third to one-half. The study found that Palo Verde and some counties in the Sacramento Valley could have the largest reductions in irrigation water deliveries and production.

Energy and Coastal Resources

Energy

Chapter 12 uses cross-sectional analysis (which compares behavior in different climates) to estimate the response of energy expenditures to climate. Separate analyses are conducted for residential and commercial users. Because the available data did not identify Californian residents, a national study was used to estimate energy sensitivity. Energy expenditures are expected to rise with warming in the summer but decline with warming in the winter. Warming in the hotter regions of the state such as the southeastern desert leads to large net increases in energy expenditures (losses). Warming in the cooler northern maritime and high alpine counties leads to much smaller effects, in some cases net reductions in energy expenditures (gains).

Statewide net costs in the residential energy sector increase as temperatures rise. Energy expenditures could increase 0.5 percent to 6 percent by 2020 and up to 9 percent by 2060. By 2100, residential energy expenditures could increase from $3.5 billion (an 8 percent increase) to $10.2 billion (a 17 percent increase) depending on the climate scenario. The increased use of energy for cooling more than offsets the reduced use of energy for heating. The higher costs account for installing and using more air-conditioning as an adaptation to climate change. Annual energy expenditures for the commercial sector would also increase with warming from $3 billion to almost $9 billion. By 2100, total energy expenditures for the state are estimated to rise from $6.7 billion for a 3°C warming to $18.9 billion (a 21 percent increase) for a 5°C warming. The damages are proportional to the size of the energy sector, so that more rapid growth leads to larger impacts. In fact, at current energy levels, the damages would be about one-third of the estimates in Table 14.1. Future energy prices are also an important factor. With higher prices, residential energy expenditures increase substantially. A larger residential energy sector increases overall expenditures because of the higher temperature sensitivity of the residential sector.

Coastal resources

Chapter 13 investigates the damages associated with sea level rise to California's low-lying developed coastal areas. The study investigates whether it is more economically efficient to protect the shoreline or to allow it to be inundated. The cost of protection must be weighed against the value of the land and buildings that would be saved. The study also examines when to protect each coastline. Although most of California's coast is cliffs, substantial portions are low-lying and highly developed, particularly in the Bay Area and in southern California. The study explores scenarios that lead to sea level rising by 33 cm, 50 cm, 67 cm, and 100 cm by 2100, which encompasses a broad range of potential sea level rise. The study then examined what would happen at selected sites along the California coast to determine the impact on the state. The study found that most of the low-lying and exposed urban coastline has sufficiently high value to justify protection by sea walls. By timing sea wall construction to anticipate sea level rise each decade, the study reduces the cost of protection substantially. The undiscounted cost of protecting vulnerable areas over the next 100 years is estimated to be approximately $700 million for a 50 cm sea level rise and $4.7 billion for a 1 m sea level rise. The discounted value is estimated to be between $46 and $123 million for a 50 cm sea level rise and between $198 and $635 million for a 1 m sea level rise. The difference between the discounted and undiscounted costs is one measure of the importance of timing. The cost for a 50 cm sea level rise by 2100 is estimated to be about $150 million/year and a 1 m sea level rise is estimated to cost $1 billion/year.

14.3   CONCLUSIONS

On the whole, the results presented here suggest that the impacts of climate change on California involve both damages and benefits. The impacts also depend on the climate scenario. The warmer and drier the scenario, the more harmful it will be. Energy and agriculture stand out as the sectors most sensitive to direct effects. Timber and agriculture are also sensitive to indirect effects through price changes. Other sectors likely to be affected are coastal and water resources. In general, the economic impacts in the energy sector dominate the impacts to market sectors in California. Because these energy impacts involve large costs, the study generally concludes that warming would, on net, result in losses for California.

In the near term, with relatively small climate changes, the damages are expected to be relatively small. However, by 2100, when California's average temperature could increase by as much as 5°C, there is the potential for large net losses across the state. The estimated market impacts are in Table 14.1. Specifically in the energy sector, the increased demand for energy to remain cool could cost the state between $7 and $19 billion. The agriculture-water sector could also have large impacts if the climate becomes very dry. In that case, agriculture could lose $1 billion per year. The direct effects on timber are expected to be small, but if timber prices fall from worldwide increases in timber supply, the indirect benefits to consumers could be quite large. A wet scenario could generate some benefits in the water sector from hydropower but a dry scenario would lead to losses. A relatively low sea level rise would lead to only minor damages but a relatively high sea level rise could cost $100 million a year by 2100. To put the estimated climate impacts in perspective, in 2003, California's economic production was $1.4 trillion (US Bureau of Economic Analysis, 2005). If California's economy grows as expected in our high growth scenario, the economy could be as large as $6–$12 trillion by 2100. The net economic effect of climate change on California from the impacts we studied would be small relative to the economy (about 0.1 percent of GDP). However, these economic impacts are not evenly distributed across the state. Some regions and sectors would bear greater losses than the rest of the economy.

For perspective, it is interesting to compare the results of this California study to a recent national study of climate change that looked at regional impacts across the United States (Mendelsohn, 2001). The national study predicted that climate changes of over 3°C would damage southern regions in the country but leave northern regions relatively unaffected on net. The California results are thus consistent with these regional predictions. This regional book found that managed systems such as water resources and agriculture have enormous potential to adapt to potential decreases in water supply. By reallocating water across users and switching crops, large reductions in water could be translated into relatively small damages. Whether there would be barriers to such adaptations is difficult to predict, but the study demonstrates the potential of adaptation to ameliorate many of the adverse impacts of climate change.

Some of the most important impacts to the state depend on how climate change affects the rest of the world. The prices in agriculture and timber depend on how climate change affects the global supply of food and wood products. If climate change has a positive effect on supply, prices will drop, negatively affecting California producers but giving California consumers a large gain. On net, global price declines will be a benefit for California because the consumer gains will outweigh the producer losses. Some important regional climate impacts are indirect effects of changes that occur elsewhere. The Intergovernmental Panel on Climate Change (IPCC) found that with a 3°C increase in mean global temperature,[2] global agricultural production would rise, resulting in lower crop prices (McCarthy et al.,

2001). Beyond that it is projected to fall, resulting in higher prices. The IPCC also projects forest production to rise (McCarthy et al., 2001).

The study also tested whether climate impacts varied with baselines. Specifically, the study tested whether high and low population growth scenarios affected climate impacts. In general, the magnitude of impacts increased with higher populations. In the case of water, higher populations increased the demand for water, making water scarcer. In scenarios where water supply declined, higher populations resulted in increased damages. In the energy model, warming costs increased with population and income as baseline energy demand increased. Increased growth puts more pressure on coastal development, forcing the coastline to be defended against rising seas. The effect of growth on agriculture land was predicted to be limited, but the effect on water might be more important. Regions within California differ in how they would be affected by climate change. Southern California faces relatively large increases in energy costs, smaller agricultural impacts, and a substantial share of the costs of sea level rise (although San Francisco Bay would also have a substantial portion of sea level rise adaptation costs). Parts of the Sacramento Valley could be vulnerable to reductions in water supply for irrigation, and the Sacramento region is at particular risk from increased flooding. In addition, northern California timber producers are at risk if climate change reduces global timber prices. On the other hand, northern and particularly coastal areas could see reduced energy costs.

Of course, all modeling efforts have limitations. All the studies in this book had to make heroic assumptions about what California would look like in the future. We examined only a limited set of climate scenarios. The cross-sectional energy and crop studies may be biased by omitted variables. The water study assumed that water managers had perfect information about future weather. Many of the studies assumed that actors would behave efficiently; this may not be the case. Information may not be freely available and existing laws may restrict efficient outcomes from occurring. These studies cannot predict the precise effects of climate change at the end of the century; the predictions are uncertain.

The study nonetheless provides important insights about climate change. The study identifies who or what gets harmed or loses versus who or what benefits, the relative degree of impacts (that is, the magnitude of gains and losses), and the potential for adaptation to ameliorate adverse impacts. Readers should not take the projections as precise predictions, but as indications of how California can be affected by climate change.

Further, many important consequences have not been valued in this book. For example, although Chapter 6 quantifies changes in biodiversity, the study does not place a value on these changes. Changes in the flow of public services that have no price are difficult to value even though they clearly matter to people. In addition to biodiversity, people may be concerned that ecosystems are shifting as a result of warming. They may be concerned that the potential for fire in the state could increase whether climate becomes drier or wetter and that changes in runoff could in turn affect wetlands and fish as well as flooding risks. People may be concerned if warming affects pollution by altering atmospheric chemistry or if it directly affects human health through heat waves. Another resource that was not valued was coastal wetlands. Higher seas are likely to inundate the wetlands, shrinking their extent. This could be particularly important in San Francisco Bay (Galbraith et al., 2002).

Of course, not everything left out of this study will increase the damages from warming. The study did not look at recreation that could well benefit from warming because of a longer summer season (which is expected to more than offset a shorter ski season (Loomis and Crespi, 1999; Mendelsohn and Markowski, 1999)). The study did not include the full suite of possible adaptations. Extensive advances have been made in energy conservation in California. If these improvements continue, they would substantially reduce the projected energy damages from warming. The height of dams could be raised to capture more of the spring melt. This would reduce flooding damages and capture more water for agriculture. Continued agronomic improvements could lead to new varieties of crops that are more heat or drought tolerant. Investments into advanced irrigation systems could yield higher returns per acre-foot of water. Many adaptations that have not yet been studied could reduce the damages to California.

The most important insight of the book is that the research shows how to do regional analyses of climate change. The study demonstrates a method for choosing baselines and climate scenarios, and it illustrates the importance of integrating across disciplines to quantitatively link climate changes to physical effects to economic impacts. This regional study provides insights into what impacts are due to changes happening in the region and what impacts are indirect effects from changes happening elsewhere in the world. The book provides a sense of the magnitude of many of the potential impacts of climate change to California. These are all contributions to our understanding of the potential impacts of climate change.

NOTES

1. The large increase in precipitation under Hadley translates into a much smaller increase in raw water availability because much of the increase is assumed to happen in winter when reservoirs are too full to capture it. Thus, most of the increase is spilled and goes out to sea. The study did not examine the effect of increasing storage potential.

2. Climate models tend to show that the average increase in temperature in California would be about the same or slightly greater than the increase in mean global temperature (Wigley, 1999).

REFERENCES

Galbraith, H., R. Jones, R. Park, J. Clough, S. Herrod-Julius, B. Harrington, and G. Page. 2002. Global climate change and sea level rise: Potential losses of inter-tidal habitat for shorebirds. *Waterbirds* **25**(2): 173–83.

Houghton, J., Y. Ding, D. Griggs, M. Noguer, P. van der Linden, X. Dai, K. Maskell, and C. Johnson (eds). 2001. *Climate Change 2001: The Scientific Basis*. Third Assessment Report of the Intergovernmental Panel on Climate Change. Cambridge University Press, Cambridge, UK.

Loomis, J. and J. Crespi. 1999. Estimated effects of climate change on selected outdoor recreation activities in the United States. In *The Impact of Climate Change on the US Economy*, R. Mendelsohn and J. Neumann (eds). Cambridge University Press, Cambridge, UK, pp. 289–314.

McCarthy, J., O. Canziani, N. Leary, D. Dokken, and K. White (eds). (2001). *Climate Change 2001: Impacts, Adaptation, and Vulnerability*. Third Assessment Report of the Intergovernmental Panel on Climate Change, Cambridge: Cambridge University Press.

Mendelsohn, R. (ed.). 2001. *Global Warming and the American Economy: A Regional Analysis*. Edward Elgar Publishing, Cheltenham, UK and Northampton, MA.

Mendelsohn, R. and M. Markowski. 1999. The impact of climate change on outdoor recreation. In *The Impact of Climate Change on the US Economy*, R. Mendelsohn and J. Neumann (eds). Cambridge University Press, Cambridge, UK, pp. 267–88.

US Bureau of Economic Analysis. 2005. Regional Economic Accounts. Gross State Product Table. Available at http://www.bea.doc.gov/bea/regional/gsp/action.cfm. Accessed 17 January 2005.

Wigley, T.M.L. 1999. *The Science of Climate Change: Global and US Perspectives*. The Pew Center on Global Climate Change, Arlington, VA.

Index

grid cells 16
logit models 16
non-spatial and spatial data 14
own-site variables 17
regulatory and administrative variables 17
Sacramento region 18
San Joaquin Valley 18
urban growth process 15

Hadley scenario (HadCM2)
agricultural land 199–205, 200
carbon budget, future 77, 78, Plate 4
carbon density 73–4, Plate 4
coastal sage scrub (CSS), potential or current distribution Plate 6
crop yields 158
energy impacts 229
flood control 172
GCM selection criteria 53, 54
Hadley 2100, water management 177
precipitation 128–9, 134, 135, Plate 1
runoff study 254
timber impacts 110, 114
vegetation classes 68–70, Plate 3
water flows, climate change effects
precipitation 128–9, 134, 135
temperature 128
water supply 255
water use, and crop yields 158
Halvorsen, R. 219
hard structure armoring, and coastal protection 235, 236
Hillel, D. 147
hillsides/steeply sloped land, baseline impact assessment 42
Household Energy Consumption and Expenditures Survey (1990) 215
hydropower, California system 256

Imperial Beach
developed shoreline in 239
and sea level rise 237
Imperial County
agricultural land, decrease in 197
baseline results 32
population projections 21
riparian areas 43
IN_CITY variable, growth model calibration 17
INC_ACCESS90 demand variable, growth model calibration 16
incremental scenarios
carbon density and fire regime 76–7
GCM selection criteria 55
vegetation classes 71–2

infill development
and forecasting method 27, 28
projected 20, 30
Inland Empire
baseline scenario results 38
farmlands 45
intrametropolitan decentralization, job growth patterns 19
inundation mapping, and sea level rise, impacts 237
IPCC (Intergovernmental Panel on Climate Change)
on future temperature changes 4
on global agricultural production 259
on sea level rise 234, 236
irrigation water use equation 151–2

Jeton, A.E. 122
job accessibility measures, forecasting method 29
job growth patterns, forecasting method 19–20
JOB_ACCESS90 demand variable, growth model calibration 16, 17

Kern County
baseline scenarios 39
riparian areas 43
Kings River (Pine Flat Dam)
runoff analysis 124, 254
precipitation 129
Knowles, N. 122

LA, *see* Los Angeles region
La Niña, and climate 50
Landis, J. 89, 91
landscape ecological diversity, biodiversity changes 93–4
Lawrence Berkeley National Laboratory, and water resources 166
Lenihan, J.M. 89
Lettenmaier, D.P. 122
liquefield petroleum gas (LPG) 221
Livesay, N. 237
local change component (LCC), job growth patterns 20
logit models, growth model calibration 16
logit regression, and probability of cooling 216
Los Angeles region
baseline scenario results 32
coastal centers 25
conservation–development conflicts 95
Highway 101 corridor 32, 40
incremental scenarios 55
infill trends 27